Green Energy and Technology

For further volumes:
http://www.springer.com/series/8059

Jarek Kurnitski
Editor

Cost Optimal and Nearly Zero-Energy Buildings (nZEB)

Definitions, Calculation Principles and Case Studies

 Springer

Editor
Jarek Kurnitski
Tallinn University of Technology
Tallinn
Estonia

ISSN 1865-3529   ISSN 1865-3537   (electronic)
ISBN 978-1-4471-6989-5   ISBN 978-1-4471-5610-9   (eBook)
DOI 10.1007/978-1-4471-5610-9
Springer London Heidelberg New York Dordrecht

Printed on acid-free paper

Springer is part of Springer Science+Business Media (www.springer.com)

Preface

Nearly zero-energy (nZEB) buildings and cost-optimal energy performance have suddenly become a widely discussed topic across Europe. How to construct these buildings, how to design them, and above all what it means are relevant questions that many building professionals and decision makers from both the public and private sector need to ask and find answers to. The current situation is historic, as the EU has to be ready for the mass construction of nZEB buildings by 2019.

Behind the scenes of this system-wide change in construction, directives on energy performance in buildings in combination with related R&D at all levels, from technology to calculation methods and regulation, have made it possible to design and construct buildings with remarkably improved energy performance. nZEB buildings are expected to use 2–3 times less energy compared to today's modern buildings, should also provide a high-quality indoor environment and long service life, and have to be easy to operate and maintain. Yet, there is still a long way to go in order to realize these ambitious goals in practice, and we hope this book represents a valuable step forward.

There are good reasons for European regulations on the energy performance of buildings: Buildings account for roughly 40 % of total primary energy use in the EU and globally, and also offer the greatest cost-effective energy saving potential compared to other sectors. Unlike the energy and transport sectors, in the building sector the technology for energy savings already exists, making rapid execution possible once the necessary skills and regulations are in place. Uniform implementation would accelerate the process, as differences in regulations complicate building design, installation and construction, as well as manufacturing and sales in the common market area.

In this book, we have collected the latest information available on nZEB buildings; the respective authors are well-versed in the preparation of European REHVA nZEB technical definitions, as well as national regulations and nZEB requirements. They present the latest information on technical definitions, system boundaries, and methodologies for energy performance calculations, as well as descriptions of technical solutions and design processes on the basis of nZEB building case studies—essential resources for all those who need to understand and/or work with the energy performance of buildings.

The authors believe that a healthy and ongoing exchange of information will help to promote more concrete and harmonized national nZEB regulations, and to find cost-effective design processes and technical solutions for future nZEB buildings.

Tallinn, July 21, 2013   Jarek Kurnitski

Contents

Introduction 1
Jarek Kurnitski

**Nearly Zero-Energy Building's (nZEB) Definitions
and Assessment Boundaries** 7
Jarek Kurnitski

**Present Energy Performance Requirements and nZEB Targets
in Some Selected Countries** 31
Jarek Kurnitski, Christian Feldmann, Per Heiselberg,
Livio Mazzarella, Igor Sartori, Karsten Voss and Åsa Wahlström

Cost Optimal Energy Performance 47
Jarek Kurnitski

Target Values for Indoor Environment in Energy-Efficient Design . . . 57
Olli Seppänen and Jarek Kurnitski

**Energy Efficiency Measures: In Different Climates
and in Architectural Competitions** 79
Panu Mustakallio and Jarek Kurnitski

**Basic Design Principles of nZEB Buildings in Scoping
and Conceptual Design** 103
Hendrik Voll, Risto Kosonen and Jarek Kurnitski

nZEB Case Studies 135
Jarek Kurnitski, Matthias Achermann, Jonas Gräslund,
Oscar Hernandez and Wim Zeiler

Authors

Matthias Achermann Amstein Walthert Geneva, Geneva, Switzerland

Jonas Gräslund Skanska Commercial Development Nordic, Stockholm, Sweden

Oscar Hernandez Elithis Groupe, Dijon, France

Risto Kosonen Halton Group, Helsinki, Finland

Jarek Kurnitski Tallinn University of Technology, Tallinn, Estonia

Panu Mustakallio Halton Group, Helsinki, Finland

Olli Seppänen Federation of European Heating, Ventilation and Air-conditioning Associations, Brussels, Belgium

Hendrik Voll Tallinn University of Technology, Tallinn, Estonia

Wim Zeiler Eindhoven University of Technology, Eindhoven, The Netherlands

Introduction

Jarek Kurnitski

Abstract Nearly zero-energy building (nZEB) requirements can be seen as a major driver in construction sector for next years, as all new buildings in EU are expected to be nearly zero buildings from 2021. In many countries, present energy performance minimum requirements have not been able to follow increasing energy prices—that has been revealed by cost optimality analyses showing the fact that requirements lag behind and are not able to provide minimal life cycle cost of construction and operation of buildings even with reasonably short life cycle periods used. These two new terms, nZEB and cost optimal energy performance, were launched by energy performance of buildings directive recast (EPBD recast) in EPBD (2010). EPBD requires that energy performance minimum requirements will be shifted to cost optimal level as a first step towards nZEB buildings. Member States have to define what nZEB for them exactly constitutes. It is easy to realize the problem that various definitions of nZEB may cause in Europe if uniformed methodology will not be used. In this book, the latest information on technical definitions, system boundaries and other methodology for energy performance calculations, as well as description of technical solutions, based on nZEB building case studies can be found. This could help all persons needing to be aware or working with energy performance of buildings.

1 EPBD Recast: Many Duties to the Member States

Energy performance of buildings directive recast came into force on 9 July 2010 [1]. The background for the directive states that buildings account for 40 % of the total energy consumption in the European Union. The sector is expanding, which

J. Kurnitski (✉)
Tallinn University of Technology, Ehitajate tee 5, 19086, Tallinn, Estonia
e-mail: Jarek.kurnitski@ttu.ee

J. Kurnitski (ed.), *Cost Optimal and Nearly Zero-Energy Buildings (nZEB)*,
Green Energy and Technology, DOI: 10.1007/978-1-4471-5610-9_1,
© Springer-Verlag London 2013

is bound to increase its energy consumption. Therefore, the reduction in energy consumption and the use of energy from renewable sources in the building sector constitute important measures, which are needed to reduce the Union's energy dependency and greenhouse gas emissions. Together with an increased use of energy from renewable sources, measures taken to reduce energy consumption in the Union would allow the Union to comply with the Kyoto Protocol, and its commitment to reduce, by 2020, the overall greenhouse gas emissions by at least 20 % below 1990 levels.

Member States shall adopt and publish, by 9 July 2012, at the latest, the laws, regulations and administrative provisions necessary to comply with most of the articles, required to be in force from 9 January 2013. According to the Directive, the Member States shall ensure that by 31 December 2020, all new buildings are nearly zero-energy buildings; and after 31 December 2018, new buildings occupied and owned by public authorities will be nearly zero-energy buildings.

In the directive, 'nearly zero-energy building' means a building that has a very high energy performance. The nearly zero or very low amount of energy required should be covered to a very significant extent by energy from renewable sources, including energy from renewable sources produced on-site or nearby. Since the Commission does not give minimum or maximum harmonized requirements, it will be up to the Member States to define what for them exactly constitutes a "very high energy performance".

National roadmaps towards nearly zero-energy buildings are needed for all Member States. Member States shall draw up national plans for increasing the number of nearly zero-energy buildings. These national plans may include targets differentiated according to the category of building. Member States shall furthermore, following the leading example of the public sector, develop policies and take measures such as the setting of targets in order to stimulate the transformation of buildings that are refurbished into nearly zero-energy buildings and inform the Commission thereof in their national plans.

The national plans shall include, inter alia, the following elements:

(a) the Member State's detailed application in practice of the definition of nearly zero-energy buildings, reflecting their national, regional or local conditions and including a numerical indicator of primary energy use expressed in kWh/m^2 per year. Primary energy factors used for the determination of the primary energy use may be based on national or regional yearly average values and may take into account relevant European standards;

(b) intermediate targets for improving the energy performance of new buildings, by 2015;

(c) information on the policies and financial or other measures adopted in the context of for the promotion of nearly zero-energy buildings, including details of national requirements and measures concerning the use of energy from renewable sources in new buildings and existing buildings undergoing major renovation.

The Commission should by 31 December 2012, and every 3 years thereafter publish a report on the progress of Member States in increasing the number of nearly zero-energy buildings. Because of some delays, the first report is published in 2013. On the basis of that report, the Commission shall develop an action plan and, if necessary, propose measures to increase the number of those buildings and encourage best practices as regards the cost-effective transformation of existing buildings into nearly zero-energy buildings.

2 Present Energy Performance Minimum Requirements

EPBD recast requirements provide a good roadmap for regulation, calculation method and technology development that is also highly needed because of not harmonized requirements in Member States. A benchmarking study on implementation on first version of EPBD 2002 by REHVA Seppänen and Goeders [2] revealed a large variation in the technical regulations of the different countries. These differences in regulations have a significant effect on the building industry and complicate manufacturing, sales, installation, design and construction of buildings in the common market area. The experience learned from the actions taken by CEN from the year 2002 to help the implementation of EPBD showed that technical development work takes time.

This is confirmed by Chap. 3 of this book, reporting the situation with national energy regulation and nZEB definitions in selected countries, showing at the same time the progress towards primary energy-based regulation, but also differences in national energy frames and calculation methods. Hopefully, Chap. 2 of this book focusing on definitions and specification of energy boundaries and calculation principles will help the experts in the Member States in defining the nearly zero-energy buildings in a uniform way in national building codes. The information provided would help in understanding the policy options and in exchanging information of most energy-efficient technical solutions for nZEB buildings.

3 nZEB: A Complex Issue of Regulatory, Methodology and Technology Challenge

At the end of the day, EU has to be ready for mass construction of nZEB buildings in 2019/2021. nZEB buildings are expected to use 2–3 times less energy compared to today's modern buildings and should be also safe and comfortable with long service life, i.e., easy to operate and maintain. It is easy to see that new technical quality is needed to achieve this ambitious target of nZEB. Technical quality is very broad term in this context. It means a new quality in regulation, in order to assure that new nZEB buildings will really provide expected energy savings and good indoor climate needed for occupants' health, comfort and productivity. For design as well as compliance assessment of cost optimal low energy or nZEB buildings, one needs relevant calculation methods and tools. Most people have understood that hand calculation methods are not enough to design energy and cost-effective buildings, and the solution will rather be in building simulation and information modelling. Questions related to definitions, regulations and calculation principles are discussed in Chaps. 2 and 3.

Chapter 4 deals with cost optimal calculations. Cost optimal policy launched by EPBD recast will instruct Member States for the first time on how to set minimum requirements and shift those away from only upfront investment cost. Because of complicated life cycle calculation procedure, the logic is that these calculations are required to test and determine cost optimality of national energy performance minimum requirements. Cost optimal calculation are not required in every construction project; however, construction clients could utilize similar type of calculations in decision making.

There is one fundamental question related to the purpose of buildings. The buildings have always been built to provide a shelter from outdoor weather. Therefore, we do not design buildings for zero energy use, but for occupants. Occupants need functional, esthetical, architectural and indoor environmental quality. The latter one can be quite easily measured and energy use depends so much on indoor climate quality that the target specification has always to include both energy and indoor climate target values. In nZEB buildings which tend to be well-insulated airtight buildings, indoor climate specification and control is even more stressed. Chapter 5 is dedicated to indoor climate parameter specification.

And last but not least we have design process and technical solutions questions of nZEB buildings. Chapter 6 helps to understand how the main energy uses are formed in different climates and which are the main energy efficiency measures in non-residential buildings. It is easy to understand that integrated design is needed to design high-performance buildings with multiple targets. Integrated design, which could be assured with careful specification of energy performance and other targets, has to be started already as earlier as in architectural competitions if such are decided to organize, discussed also in Chap. 6.

Chapter 7 will continue with important design issues in scoping and conceptual design phase. These early stages show the major differences compared with conventional design process and therefore require major effort in order to control that design streams are going in right direction so that specified energy performance targets can be achieved with cost-effective manner.

One can be sure that majority of cost-effective solutions for nZEB buildings are not yet developed. In the other hand, all required technology exists today, and the challenge is in effective design process management and development of cost-effective mass production applications. Five nZEB office building case studies reported in Chap. 8 from across the Europe reveal that the technology do exists and with good skills nZEB buildings can be built, and in some cases, this has been even cost effective.

References

1. EPBD (2010) Directive 2010/31/EU of the European parliament and of the council of 19 May 2010 on the energy performance of buildings (recast). http://ec.europa.eu/energy/efficiency/buildings/buildings_en.htm, http://eur-lex.europa.eu/JOHtml.do?uri=OJ:L:2010:153:SOM:EN:HTML
2. Seppänen O, Goeders G (2010) Benchmarking regulations on energy efficiency of buildings. Executive summary. Federation of European Heating, Ventilation and Air-conditioning Associations—REHVA, 5 May 2010

Nearly Zero-Energy Building's (nZEB) Definitions and Assessment Boundaries

Jarek Kurnitski

Abstract How to define nearly zero-energy buildings? This is one major question for member states when implementing EPBD directive. The directive requires that energy performance of buildings, as well as the definition of nearly zero-energy buildings, should be expressed with a numerical indicator of primary energy in kWh/m^2 per year. Therefore, if a national energy frame (energy calculation methodology and minimum requirements) is not based on primary energy, first a new energy frame/methodology has to be developed and implemented in a building code in order to be able to implement the directive. This could be a major effort in countries where minimum energy performance requirements have been based on the requirements of building components or on energy need or on delivered energy. In this chapter, energy flows needed for the primary energy indicator calculation are described based on REHVA and CEN definitions, Kurnitski REHVA Report No 4, (2013), prEN 15603 (2013). A specific issue of nZEB buildings is accounting the positive effect of on-site and nearby renewable energy production, which needs to be included in the energy frame. Energy frame is described with system boundaries for each energy calculation step, starting from the energy need to the final system boundary of the delivered and exported energy, which allows us to calculate primary energy indicator and renewable energy contribution with national primary energy factors. Because of complicated definitions and system boundaries, calculation examples for all main cases are provided.

J. Kurnitski (✉)
Tallinn University of Technology, Ehitajate tee 5, 19086, Tallinn, Estonia
e-mail: Jarek.kurnitski@ttu.ee

J. Kurnitski (ed.), *Cost Optimal and Nearly Zero-Energy Buildings (nZEB)*,
Green Energy and Technology, DOI: 10.1007/978-1-4471-5610-9_2,
© Springer-Verlag London 2013

1 Technical Definition for Nearly Zero-Energy Buildings

Technical definitions and energy calculation principles for nZEB are needed to clarify the exact technical meaning of EPBD recast [1] requirements in order to support uniformed national implementation. EPBD recast requires nearly nZEB buildings, defined as buildings with **a very high energy performance** and where energy need is covered to **a very significant extent by energy from renewable sources** (original wording given in Sect. 2.1). Since EPBD recast does not give minimum or maximum harmonized requirements as well as details of energy performance calculation framework, it will be up to the member states to define what "a very high energy performance" and "to a very significant extent by energy from renewable sources" for them exactly constitute.

EPBD recast requires the evaluation of the cost optimality of current national minimum requirements by March 2013 with the methodology called "delegated Regulation supplementing Directive 2010/31/EU" published in March 21, 2012. It is recommended to use the same system boundary and energy calculation framework for both cost-optimal and nZEB energy calculations. Cost-optimal performance level means the energy performance in terms of primary energy, leading to minimum life-cycle cost. This cost-optimal policy launched by EPBD recast will instruct MS to shift minimum requirements to cost-optimal energy performance level. Cost-optimal policy does not say that nZEB has to be cost optimal, because nZEB is another, next political target established by EPBD. According to current understanding, nZEB is not cost-effective yet; however, this may depend on available incentives. Therefore, these both requirements (cost optimal and nZEB) will have to be reconciled so that a smooth transaction from cost-optimal requirements to nearly zero-energy buildings could be guaranteed in near future. Currently, it is suggested to define nZEB performance level rather through the bases of reasonably achievable technical solutions instead of cost optimal bases, which may be the situation in the future. Cost-optimal calculations are straightforward for the solutions with well-established costs that do not apply for renewable technologies where rapid developments are expected to make such calculations uncertain.

The following definition includes the system boundary specifying how to define the various energy flows and description of energy calculation framework which will also affect the performance levels of nZEB building definitions. This guidance will help the experts in the member states in defining the nearly zero-energy buildings in a uniform way.

Nearly zero-energy building definition shall be based on delivered and exported energy according to EPBD recast [1] and prEN 15603 [9]. The basic energy balance of the delivered and exported energy and system boundaries for the primary and renewable energy calculations are shown in Figs. 1 and 2 and described with detailed system boundary definitions in Sect. 2.2, Figs. 5, 6, 7, and 8. These system boundary definitions apply for a single building or for sites with multiple buildings with or without nearby production as discussed in Sect. 2.2. Principles of EPBD

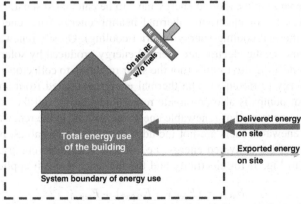

Building site = system boundary of delivered and exported energyon site

Fig. 1 System boundaries for on-site assessment (nearby production not linked to the building) for nearly zero-energy building definition, connecting a building to energy networks and using on-site renewable energy (*RE*) sources. System boundary of energy use of building technical systems follows outer surface of the building in this simplified figure; system boundary of delivered and exported energy on site is shown with *dashed line*. In the case of nearby production, the nearby system boundary will be added as shown in Fig. 3 and explained with detailed system boundaries in Sect. 2.2

Fig. 2 An example of an all-electrical building explaining the use of Eqs.1 and 2

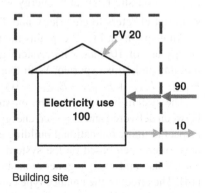

Building site

recast and the description of the assessment boundary given in prEN 15603 [9] are followed. Total energy use of the building suggests that all energy used in buildings will be accounted. According to EPBD recast (definition shown in Sect. 2.1), all components of the energy use are mandatory except the energy use of appliances (households, elevators/escalators, and outlets) which may or may not be included. With the inclusion of appliances, energy use in the buildings includes energy used for heating, cooling, ventilation, and hot water lighting appliances. Taking into account appliances is also needed for the calculation of exported energy or to analyze load matching and grid interaction.

   Delivered energy and exported energy have to be calculated separately for each
energy carrier, i.e., for electricity, thermal heating energy (fuel energy, district
heating), and thermal cooling energy (district cooling). On-site renewable energy
without fuels means the electric and thermal energy produced by solar collectors,
PV, wind turbine, or hydro turbine (not the solar radiation to collectors or panels or
the kinetic energy to turbines). The thermal energy extracted from ambient heat
sources by heat pumps is also an on-site renewable energy, and the ambient heat
exchangers may be treated as renewable energy generators in the renewable energy
calculation. Renewable fuels are not included in on-site renewables, but they are
renewable part of the delivered energy, i.e., off-site renewable energy.
   According to Fig. 1, for electricity and thermal energy use it applies:

$$E_{us,el} = \left(E_{del,el} - E_{exp,el} \right) + E_{ren,el} \tag{1}$$

and

$$E_{us,T} = \left(E_{del,T} - E_{exp,T} \right) + E_{ren,T} \tag{2}$$

where
E_{us} is total energy use kWh/(a);
E_{del} is delivered energy on site (kWh/a);
E_{exp} is exported energy on site (kWh/a); and
E_{ren} is on-site renewable energy without fuels (kWh/a);
subscript el refers to electricity and T to thermal energy.
   An example in Fig. 2 explains the use of Eq. 1. An all-electrical building with
energy use of 100 has a PV system generating 20, from which 10 is used in the
building and 10 is exported. With these values, delivered energy on site becomes:
$E_{del,el} = E_{us,el} + E_{exp,el} - E_{ren,el} = 100 + 10 - 20 = 90$.
   The system boundary for the on-site assessment, shown in Fig. 1 (the balance
between delivered and exported energy on site), draws a remarkable difference with
respect to many old national building energy codes and calculation procedures that
have often been based on the system boundary of energy use (a balance between
energy use and energy generation, but exported energy is neglected/not defined
[14]. The effect of the balance type can be seen from example in Fig. 2. Energy use
is 100, delivered energy on site is 90 and delivered energy minus exported energy
on site is $90 - 10 = 80$; old national definitions have been typically been based on
first two options. As discussed in Sartori et al. [16] and Voss et al. [15], the balance
based on delivered and exported energy creates the need to calculate on-site generation used in a building and exported (i.e., in the case of Fig. 2 to be able to
calculate that 10 of PV generation is used in the building and 10 is exported).
   Generally, because of on-site energy generation (load match) and dynamics in
energy use (cooling, intermittent operation), **an hourly energy calculation
(energy simulation)** is needed to calculate the delivered and exported energy. This
calculation will need relevant occupancy and operation profiles as a part of
standard energy calculation input data including also hourly test reference year.

In many building codes, such hourly profiles and test reference years are not yet available and will need to be developed in order to be capable of nZEB compliance assessment. It should be also noticed that the sum of the results of single buildings, calculated with the standard input data, does not represent energy use pattern of the corresponding group of buildings, because the patterns of energy use and production are stochastically distributed due to actual occupancy and weather data in reality.

If an hourly energy simulation is seen as too complicated for residential buildings, there are some possibilities to use simplified methods. It is shown [12] that if there is no load match and if dynamics is limited, that is the case in residential buildings without on-site energy generation and cooling, monthly energy calculation provides reasonably similar results to energy simulation. To simplify the load match calculation in residential buildings, one possibility may be to define standardized on-site generation fractions for various types of generation technologies (e.g., for PV, CHP, and wind in houses and apartment buildings). For example, the method based on tabulated specific heat loss values and fixed building technical systems, used as alternative method in Estonian regulation for compliance assessment of houses, has shown accuracy not worser than 14 % for the cases with practical relevance [13].

In order to be able to take into account a new nearby renewable energy production capacity contractually linked to the building and providing the real addition in the renewable capacity to the grid or district heating or cooling mix in connection with construction/development of the building(s), the system boundary of Fig. 1 has to be extended. To calculate delivered and exported energy nearby, the energy flows of nearby production plant contractually linked to building should be added/subtracted to the delivered and exported energy flows on site [7], Fig. 3. Prerequisite to apply this nearby assessment is the availability of national legislation allowing to allocate such new capacity to the building/development with a long-term contract and assuring that the investment on that new capacity will lead to a real addition to the grid or district heating or cooling mix.

Primary energy indicator (called often also as primary energy rating) sums up all delivered and exported energy (electricity, district heat/cooling, fuels) into a single indicator with national primary energy factors. This primary energy indicator can be used to define the performance level of nearly zero-energy building. In a similar fashion, numeric indicator of CO_2 emission may be calculated with CO_2 emission coefficients. CO_2 indicator provides additional information about the consequences of energy use, in terms of CO_2 emitted to atmosphere in energy production. A complementary indicator is the share of renewable energy in the energy use of a building, the renewable energy ratio, that is discussed in Sect. 2.3.

Primary energy and primary energy indicator are calculated from delivered and exported energy as:

$$E_{P,\mathrm{nren}} = \sum_i \left(E_{\mathrm{del},i} f_{\mathrm{del,nren},i} \right) - \sum_i \left(E_{\mathrm{exp},i} f_{\mathrm{exp,nren},i} \right) \tag{3}$$

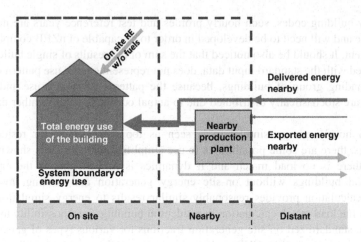

Fig. 3 Nearby assessment boundary to be used in the case of nearby energy production linked contractually to the building. Compared to on-site assessment boundary, delivered and exported energy flows on site are replaced by delivered and exported energy flows nearby

$$EP_P = \frac{E_{P,\text{nren}}}{A_{\text{net}}} \tag{4}$$

where

EP_P	is the primary energy indicator (kWh/(m^2 a));
$E_{P,\text{nren}}$	is the non-renewable primary energy (kWh/a);
$E_{\text{del},i}$	is the delivered energy on site or nearby (kWh/a) for energy carrier i;
$E_{\text{exp},i}$	is the exported energy on site or nearby (kWh/a) for energy carrier i;
$f_{\text{del,nren},i}$	is the non-renewable primary energy factor ($-$) for the delivered energy carrier i;
$f_{\text{exp,nren},i}$	is the non-renewable primary energy factor ($-$) of the delivered energy compensated by the exported energy for energy carrier i, which is by default equal to the factor of the delivered energy, if not nationally defined in other way; and
A_{net}	is useful floor area (m^2) calculated according to national definition.

In the case of nearby production contractually linked to the building, delivered and exported energy flows nearby are used; otherwise, delivered and exported energy flows on site are used. In the case of district heating or cooling network not specially linked to the building, delivered and exported energy flows on site are used, and the primary energy factor of the network mix is used (in such a case, a district heating or cooling is treated as a delivered energy flow, despite location being nearby). Contractually linked to the building means that new renewable production capacity is constructed, and in such a case, delivered and exported energy flows nearby allow to use a primary energy factor specific to this new production capacity. Generally, this requires the availability of legislation allowing us to connect new renewable generation capacity to a building for a long term (and to be treated equally with on-site production).

CO_2 emissions of energy use is calculated from delivered and exported energy as:

$$EP_{CO_2} = \frac{m_{CO_2}}{A_{net}} = \frac{\sum_i \left(E_{del,i} K_{del,i}\right) - \sum_i \left(E_{exp,i} K_{exp,i}\right)}{A_{net}} \tag{5}$$

where

EP_{CO2} is the specific CO_2 emission $(gCO_2/(m^2 \text{ a}))$;

m_{CO2} is the CO_2 emission (gCO_2/a);

$E_{del,i}$ is the delivered energy on site or nearby (kWh/a) for energy carrier i;

$E_{exp,i}$ is the exported energy on site or nearby (kWh/a) for energy carrier i;

$K_{del,i}$ is the CO_2 emission coefficient (gCO_2/kWh) for the delivered energy carrier i;

$K_{exp,i}$ is the CO_2 emission coefficient (gCO_2/kWh) for the exported energy carrier i, which may or may not be equal to the factor of the delivered energy, depending on national definition; and

A_{net} is useful floor area (m^2) calculated according to national definition.

Primary energy factors are by the definition in Fig. 1 non-renewable primary energy factors. National primary energy factors are to be used according to EPBD recast; in most of member states, these factors are already based on non-renewable primary energy considerations, however, often including energy policy considerations. For renewable fuels, non-renewable primary energy factor may include indirect effects such as transportation, etc. With these considerations, the factor may still remain as low as 0.1–0.2, which may be seen as easily leading to waste of renewable fuels as well as conflicting with energy prices. This has led to energy policy factors of about 0.5 or higher in many countries.

In order to be a sound definition, nearly zero-energy building defined through primary energy indicator shall refer to a specified energy calculation framework, including the following:

• system boundaries of energy use, renewable energy use, and delivered and exported energy (Figs. 1, 3, 5, 6, 7, 8);
• standard energy calculation input data [10];
• test reference year to be used in energy calculations [11];
• primary energy factors for energy carriers (default values in prEN 15603 [9]);
• energy calculation rules and methods for energy need and system calculations, covered in relevant EPBD standards.

which all affect calculated or measured primary energy indicator.

Net-zero-energy building definition has an exact performance level of 0 kWh/$(m^2 \text{ a})$ non-renewable primary energy. If primary energy is <0, the building is a plus-energy building. The definition refers to annual balance of primary energy [16] calculated from delivered and exported energy with Eq. 3. In some countries, a balance period of one month is used for PV electricity. In such a case, the maximum amount of exported PV electricity that can be taken into account in the energy balance is limited to the amount of the delivered electricity each month, leading to the situation that in summer months, all exported PV electricity cannot be taken into account. The performance level of "nearly" zero energy is a subject of national decision taking into account the following:

• Technically reasonably achievable level of primary energy use;
• How many % of the primary energy is covered by renewable sources;
• Available financial incentives for renewable energy or energy efficiency measures;
• Cost implications and ambition level of the definition.

The following definitions have been prepared for uniform EPBD recast implementation [7]:

Net-zero-energy building (ZEB)
Non-renewable primary energy of 0 kWh/(m^2 a).
nearly zero-energy building (nZEB)

Technically and reasonably achievable national energy use of >0 kWh/(m^2 a) but no more than a national limit value of non-renewable primary energy is achieved with a combination of best practice energy efficiency measures and renewable energy technologies which may or may not be cost optimal.

Note 1 "reasonably achievable" means by comparing with national energy use benchmarks appropriate to the activities served by the building or any other metric that is deemed appropriate by each EU Member State.

Note 2 The commission has established a comparative methodology framework for the calculation of cost-optimal levels (cost optimal).

Note 3 Renewable energy technologies needed in nearly zero-energy buildings may or may not be cost-effective, depending on available national financial incentives.

For the national definition of nearly net-zero-energy buildings, the performance levels of E-values should be specified for each building type, at least for those listed in EPBD recast:

(a) single-family houses of different types;
(b) apartment blocks;
(c) offices;
(d) educational buildings;
(e) hospitals;
(f) hotels and restaurants;
(g) sports facilities;
(h) wholesale and retail trade services buildings;
(i) other types of energy-consuming buildings.

The use of standard energy calculation input data and energy calculation rules makes it possible to compare objectively the energy performance of different buildings for compliance assessment purposes within the building types listed. In actual operation, buildings can be operated and used very differently within the same building type. But, as all of these buildings are calculated with the same input data and calculation rules, the results remain reliable for the compliance assessment. Standard energy calculation input data are not suitable for the assessment of actual energy use in a specific building. If energy performance certificates include the assessment of actual energy use, inclusion of actual building operation data as well as actual climate data and in some cases more detailed definition of building types would be needed for better accuracy.

2 Detailed System Boundaries for Delivered and Exported and Renewable Energy Calculation

2.1 Definitions Related to the System Boundaries

For any energy performance indicator as well as for low-energy or zero-energy building definition, it would be necessary to specify which energy flows are included in the definition and which ones are not. Usually, all energy used in the buildings is recommended to be taken into account, but EPBD recast allows to exclude electrical energy use of occupant appliances. Such energy flow specification is linked to system boundaries, and it provides a general framework for energy indicators. According to EPBD recast, energy performance is defined as (article 2):

> 'energy performance of a building' means the calculated or measured amount of energy needed to meet the energy demand associated with a typical use of the building, which includes, **inter alia, energy used for heating, cooling, ventilation, hot water and lighting**.

This energy performance definition helps to understand the EPBD recast definition for nearly zero-energy building (nZEB):

> 'nearly zero-energy building' means a building that has a **very high energy performance**, as determined in accordance with Annex I. The nearly zero or very low amount of energy required should be covered to a **very significant extent by energy from renewable sources**, including energy from renewable sources produced on-site or nearby.

According to these EPBD recast definitions, electricity for households, transport (elevators/escalators), and appliances/outlets are not obligatory to be included. All other major energy flows are mandatory to be included. Therefore, it is upon national decision to decide whether to take into account electricity for households and outlets or not. If taken into account, the measured and calculated energy use include the same energy flows, Fig. 4.

EPBD Recast Article 9—Nearly zero-energy buildings, 3. (a) requires that the national plans shall include:

Fig. 4 In the measured ratings, typically all energy flows are included as measured. In the calculated energy ratings, electricity for households and outlets ("others") may or may not be included

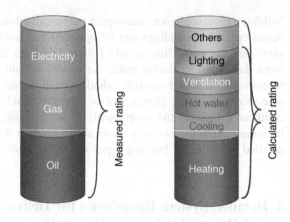

the Member State's detailed application in practice of the definition of nearly zero-energy buildings, reflecting their national, regional or local conditions, and including a **numerical indicator of primary energy use expressed in kWh/m^2 per year**. Primary energy factors used for the determination of the primary energy use may be based on national or regional yearly average values and may take into account relevant European standards;

EPBD recast, Annex I, states common general framework for the calculation of energy performance of buildings. In this framework, it is said that:

> The energy performance of a building shall be expressed in a transparent manner and shall include an **energy performance indicator and a numeric indicator of primary energy use**, based on primary energy factors per energy carrier, which may be based on national or regional annual weighted averages or a specific value for on-site production.

Therefore, EPBD states that it is possible to add in parallel other indicators, but all member states must have the **primary energy indicator in kWh/m^2 per year**. "The cost-optimal delegated Regulation supplementing Directive 2010/31/EU" [4, 5] published in March 21, 2012, requires the use of non-renewable primary energy in the cost optimality assessment. This forms a strong indication that national energy performance requirements have to be based in the future on non-renewable primary energy indicator; otherwise, the cost optimality assessment would not be possible.

In the Annex I, it is also referred to the use of relevant European standards:

> The methodology for calculating the energy performance of buildings should take into account European standards and shall be consistent with relevant Union legislation, including 2009/28/EC [2, 3].

EN 15603:2003 [8] (currently under revision, prEN 15603:2013 [9] available) specifies general framework for the assessment of energy performance of buildings that is much more detailed compared to EPBD recast definitions. This is used for detailed system boundary specification discussed in the next chapter.

2.2 Detailed System Boundary for Energy Calculation and nZEB Definition

REHVA has developed in cooperation with CEN a set of detailed system boundaries, which have been based on REHVA nZEB definition [6] and prEN 15603:2013 [9]. As stated in EPBD recast, the positive influence of renewable energy produced on site is taken into account so that it reduces the amount of delivered energy needed and may be exported if it cannot be used in the building (i.e., on site production is not considered as part of delivered energy on site), Fig. 5.

The energy calculation direction is from the energy need of rooms of a building to the energy use of technical systems, which is covered by on-site renewable energy production and delivered energy. On-site renewable and cogenerated energy productions that mismatch the building energy use are exported. This applies also for thermal energy if possible to export. The box of "building needs" refers to rooms (thermal zones) in a building, and the box of energy use refers to building technical systems showing that these can be partially located out of the building. The outer system boundary line follows the building site boundary, and

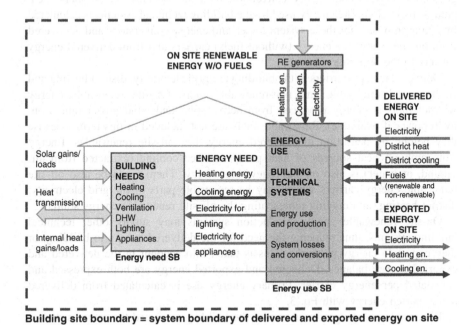

Fig. 5 Three system boundaries (*SB*) for on-site assessment (nearby production not linked to the building), for energy need, energy use, and delivered and exported energy calculation. System boundary of energy use applies also for renewable energy ratio calculation with inclusion of RE from geo-, aero-, and hydro-thermal energy sources of heat pumps and free cooling as shown in Fig. 8 [7]

the connection points to energy networks are shown by double arrows of delivered and exported energy.

Building needs in Fig. 5 represent thermal and electrical energy needs in a building for heating, cooling, ventilation, domestic hot water, lighting, and appliances. Energy need for space heating is caused by heat losses and heating of supply air from supply air temperature to room temperature (i.e., thermal energy without any system losses needed to keep a room temperature). Energy need for space heating is calculated with solar and internal heat gains which reduce energy need. Energy need for supply air heating is calculated with heat recovery to heat supply air from the temperature after heat recovery to supply air temperature. In the case of exhaust ventilation or no heat recovery, energy need for heating ventilation air equals to energy need to heat ventilation air from outdoor to room temperature. Fan electricity for ventilation is not an energy need, but belongs to the energy use of ventilation system. For the domestic hot water, the energy need equals to the thermal energy to heat cold water to hot water temperature. For the lighting and appliances, energy needed/used is electrical energy.

Building technical systems supply the amount of energy needed for heating and cooling and electrical energy. To supply these energy needs, building technical systems have typically some system losses (emission, distribution, storage, and generation losses) and energy conversion in some systems (i.e., boilers, heat pumps, fuel cells). The energy used by the building technical systems is calculated by taking into account these system losses and energy conversions and is covered from on-site renewable energy (without fuels) carriers and from delivered energy carriers to the building.

Delivered energy carriers to the building are grid electricity, district heating and cooling, and renewable and non-renewable fuels. On-site renewable energy without fuels is energy produced from active solar and wind power (and from hydropower if available). Renewable fuels are not included in this term, because they belong by definition to delivered energy, i.e., off-site renewables. Energy from ambient heat sources of heat pumps or free cooling (extracted from air, ground, or water) is also on-site renewable energy. There could be also off-site renewable energy carriers which may be renewable parts of the grid electricity, district heating, and cooling representing the off-site renewable energy sources.

On-site renewable energy production systems may supply other technical building systems, thus reducing the need for the delivered energy to building, or may export to energy networks. This is taken into account in the delivered and exported energy balance. Delivered and exported energy are both expressed and calculated per energy carrier. Primary energy use is calculated from delivered and exported energy with Eq. 3.

Renewable energy produced nearby the building is treated the same as distant but with possible different primary energy factors for energy delivered/exported from/to nearby versus energy delivered/exported from/to distance (macro-infrastructure, grid). Nearby plants can be taken into account as follows:

- With a different primary energy factor than that of the grid or the network mix if nearby production is linked to the building;
- With the primary energy factor of the network mix (for common clients of district heating or cooling);
- With the system boundary extension for a site with multiple buildings and site energy center.

Renewable fuels and energy sources used in district heating or cooling production (any district heating or cooling network may be treated as decentralized production nearby) will reduce the non-renewable primary energy factor of district heating or cooling mix. Therefore, the system boundary in Fig. 5 applies for district heating and cooling mix as it is, which means that nearby production is taken into account by the primary energy factor of the network mix. For example, if 20 % of the fuel energy in district heat production is renewable, primary energy factor of 0.8 will be used (in practice, the factor has to be calculated with network losses and CHP production effects if exists).

Another primary energy factor for nearby production than that of the grid or the network mix may be used only if the building owner makes a long-term investment on new nearby renewable energy production capacity. This applies for the cases where energy produced nearby is distributed through the grid or district heating or cooling or any other specific network. Prerequisite to use the primary energy factor of the specific nearby renewable energy plant is that the investment will lead to a real addition to the grid or district heating or cooling mix, and this is caused by and allocated to the building/development for a long term. For such cases, the nearby energy flows are treated as shown in Fig. 6. To calculate the primary energy indicator with nearby production, delivered and exported energy nearby is used in Eq. 3 instead of delivered and exported energy on site.

For the sites with multiple buildings and site energy centers, the system boundary in Fig. 5 has to be extended so that it covers the entire site with multiple buildings and decentralized production as shown in Fig. 7. Such situation typically applies for multiple buildings with a single owner, e.g., universities, hospitals, and military/government establishments. Buildings and site energy center may have on-site energy production and energy exchange between buildings. To calculate primary energy or CO_2 emissions, total useful floor area of all buildings on the site have to be used in Eqs. 3 and 5:

$$A_{net} = \sum_{i=1}^{n} A_{net,i} \qquad (6)$$

where n is the number of buildings.

Some countries accept energy performance certificates for the multiple buildings in the same site, but some countries require certificates for every building. In the latter case, energy flows are to be split for each building. District heating and

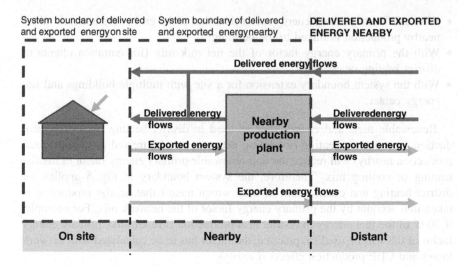

Fig. 6 System boundary for nearby production plants contractually linked to the building (typically, some share of the capacity or production of the plant can be linked to the building)

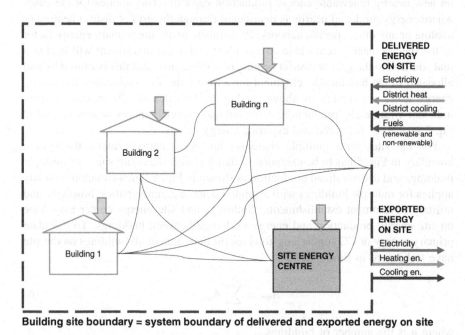

Fig. 7 System boundary for building sites with multiple buildings

cooling or any other nearby renewable energy production out of the site has to be taken into account in the same way as for a single building (with the factors of the mix or as shown in Fig. 6).

2.3 Renewable Energy Ratio Calculation

The share of renewable energy use, renewable energy ratio RER is a complementary indicator to the primary energy indicator, because renewable energy solutions are one measure to reduce non-renewable primary energy—the main target of EPBD. RER assessment is directly not required by EPBD, but as significant amount of energy use of nZEB buildings has to come from renewable energy produced on site or nearby (see original wording in Sect. 2.1), calculation equation is needed for RER. EPBD operates with primary energy that is the reason why the RER is also calculated relative to the total primary energy according to [7, 9]. Renewable energy share is also mentioned in RES directive, where primary energy is not addressed. Therefore, according to RES directive, the renewable energy share is possible to calculate with alternative approaches, which could be based on energy needs or energy uses, but are not discussed in [7, 9].

In order to calculate RER, all renewable energy sources have to be accounted for. These include solar thermal, solar electricity, wind and hydroelectricity, renewable energy captured from ambient heat sources by heat pumps and free cooling, renewable fuels, and off-site renewable energy. Ambient heat sources of heat pumps and free cooling are to be included to the renewable energy use system boundary, because in RER calculation, heat pumps and free cooling are not only taken into account with delivered energy calculation based on COP, but also taken into account by the extracted energy from ambient heat sources. Renewable energy use system boundary is shown in Fig. 8. It is important to notice that the passive solar energy belongs to energy need system boundary, not to energy use and renewable energy use system boundaries. Therefore, passive solar is not accounted in the RER calculation (also excluded by the RES directive).

The renewable energy ratio is calculated relative to all energy use in the building, in terms of total primary energy [7, 9]. It is taken into account that exported energy compensates delivered energy. By default, it is considered that the exported energy compensates the grid mix or in the case of thermal energy, the district heating or cooling network mix.

For on site and nearby renewable energy, the total primary energy factor is 1.0. For delivered energy, non-renewable and total primary energy factors are needed. Total primary energy-based RER equation is the following [7]:

$$\mathrm{RER}_P = \frac{\sum_i E_{\mathrm{ren},i} + \sum_i \left(\left(f_{\mathrm{del,tot},i} - f_{\mathrm{del,nren},i} \right) E_{\mathrm{del},i} \right)}{\sum_i E_{\mathrm{ren},i} + \sum_i \left(E_{\mathrm{del},i} f_{\mathrm{del,tot},i} \right) - \sum_i \left(E_{\mathrm{exp},i} f_{\mathrm{exp,tot},i} \right)} \qquad (7)$$

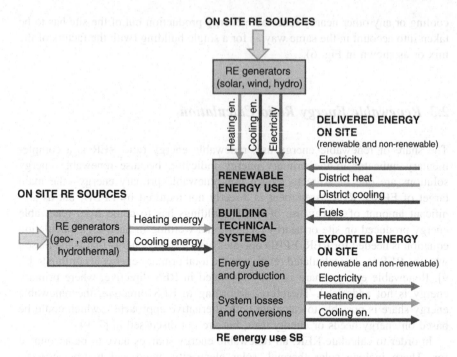

Fig. 8 Renewable energy use system boundary for renewable energy ratio RER calculation. In addition to energy flows shown in Fig. 5, renewable thermal energy from ambient heat pump and free cooling sources (heat exchangers) is accounted

where

RER_P	is the renewable energy ratio based on the total primary energy;
$E_{ren,i}$	is the renewable energy produced on site or nearby for energy carrier i, kWh/a;
$f_{del,tot,i}$	is the total primary energy factor (−) for the delivered energy carrier i;
$f_{del,nren,i}$	is the non-renewable primary energy factor (−) for the delivered energy carrier i;
$f_{exp,tot,i}$	is the total primary energy factor (−) of the delivered energy compensated by the exported energy for energy carrier i;
$E_{del,i}$	is the delivered energy on site or nearby for energy carrier i, kWh/a;
$E_{exp,i}$	is the exported energy on site or nearby for energy carrier i, kWh/a.

Term $E_{ren,site}$ represents the renewable primary energy produced on site, and they have the total primary energy factor of 1.0 and the non-renewable primary energy factor of 0. In the case of nearby energy production, the terms of the delivered energy on site and exported energy on site are replaced by the terms of the delivered energy nearby and exported energy nearby.

In addition to the calculation of the RER relative to all energy use in the building (Eq. 7), the calculation of RER is also possible in relation to specific service as heating or cooling or domestic hot water, etc. depending on national specification.

3 Calculation Examples

The following calculation examples, developed by REHVA [7], will explain the calculation logic of energy flows and different system boundaries in order to be able to calculate the primary energy indicator and renewable energy contribution. All calculation examples start from energy needs that are provided as an input data. In practice, energy needs may be calculated with energy simulation tools (hourly calculation, generally required for nZEB buildings) or national calculation tools (typically based on monthly methods, providing more indicative results). From energy needs, all calculation steps are described in examples. Because of complicated nature of system loss calculations of building technical systems, distribution and emission losses are neglected in all examples. This means that generation losses (boiler efficiency, seasonal performance factor of heat pump, etc.) are taken into account, but distribution and emission losses (heat losses of the heating or cooling pipework and losses from losses from heat emitters or cooling devices in rooms) are not taken into account to keep examples enough transparent and easy to follow. In reality, these losses are to be taken into account, and they slightly increase energy use.

3.1 Low-Energy Detached House

Consider a detached house located in Helsinki with net area 150 m² and the following annual energy needs:

- 7,200 kWh [48.0 kWh/(m² a)] energy need for heating (including ventilation and DHW);
- 1,600 kWh [10.7 kWh/(m² a)] energy need for cooling;
- 1,050 kWh [7.0 kWh/(m² a)] electricity for lighting;
- 2,400 kWh [16.0 kWh/(m² a)] electricity for appliances.

In this building, solar thermal energy provides 2,100 kWh/a [14.0 kWh/(m² a)] domestic hot water. The rest of the heating need is supplied with ground source heat pump system, which has the seasonal performance factor of 3.2. To simplify the calculation, emission, distribution, and storage losses of the heating system are neglected in this example. Only generation losses, which are included in the seasonal performance factor, are taken into account.

Energy calculation results are shown in Fig. 9. First, on-site thermal energy of 14.0 kWh/(m^2 a) is reduced from the energy need of 48.0 kWh/(m^2 a). Heat pump thus produces 34.0 kWh/(m^2 a) thermal energy with an electrical energy input of 10.6 kWh/(m^2 a). The seasonal performance factor includes circulation pumps of the heating system and the ground loop. It is considered that the ground loop is utilized for cooling, so that the circulation pump operation for cooling and the fan energy of the fan coil is 1.8 kWh/(m^2 a). Delivered electricity is 40.4 kWh/(m^2 a), and there is no exported energy (no on site production). For the grid electricity, it is assumed that it is 100 % non-renewable with the total and non-renewable primary energy factor of 2.5.

3.2 nZEB Office Building

Consider an office building located in Paris with following annual energy needs [all values are specific values in kWh/(m^2 a)]:

- 3.8 kWh/(m^2 a) energy need for heating (space heating, supply air heating, and DHW);
- 11.9 kWh/(m^2 a) energy need for cooling;
- 21.5 kWh/(m^2 a) electricity for appliances;
- 10.0 kWh/(m^2 a) electricity for lighting.

Breakdown of the energy need is shown in Fig. 10.

The building has a gas boiler for heating with a seasonal efficiency of 90 %. For cooling, free cooling from boreholes (about 1/3 of the need) is used, and the rest is covered with mechanical cooling. For borehole cooling, seasonal energy efficiency ratio of 10 is used and for mechanical cooling 3.5. To simplify the calculation, emission and distribution losses of the heating and cooling systems are neglected in this example. Ventilation system with a specific fan power of 1.2 kW/(m^3/s) and the circulation pump of the heating system will use 5.6 kWh/(m^2 a) electricity. There is installed a solar PV system providing 15.0 kWh/(m^2 a), from which 6.0 is utilized by the building and 9.0 is exported to the grid.

Energy calculation results are shown in Fig. 10, in the building technical system box. Gas boiler with 90 % efficiency results in 4.2 kWh/(m^2 a) fuel energy. Electricity use of the cooling system is calculated with seasonal energy efficiency ratios 10 and 3.5, respectively. Electricity use of free cooling, mechanical cooling, ventilation, lighting, and appliances is 39.8 kWh/(m^2 a). Solar electricity of 6.0 kWh/(m^2 a) used in the building reduces the delivered electricity to 33.8 kWh/(m^2 a). The rest of PV electricity, 9.0 kWh/(m^2 a), is exported. The delivered fuel energy (caloric value of delivered natural gas) is 4.2 kWh/(m^2 a).

In this example, it is considered that 20 % of the grid electricity is from renewable sources with the non-renewable primary energy factor of 0 and the total primary energy factor of 1.0. For the rest 80 % of the grid electricity, the total and

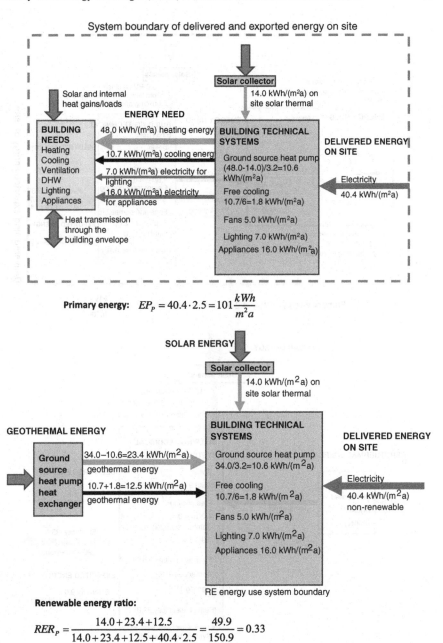

Primary energy: $EP_p = 40.4 \cdot 2.5 = 101 \dfrac{kWh}{m^2 a}$

Renewable energy ratio:

$$RER_p = \frac{14.0 + 23.4 + 12.5}{14.0 + 23.4 + 12.5 + 40.4 \cdot 2.5} = \frac{49.9}{150.9} = 0.33$$

Fig. 9 Calculation example of the primary energy and renewable energy ratio in a detached house

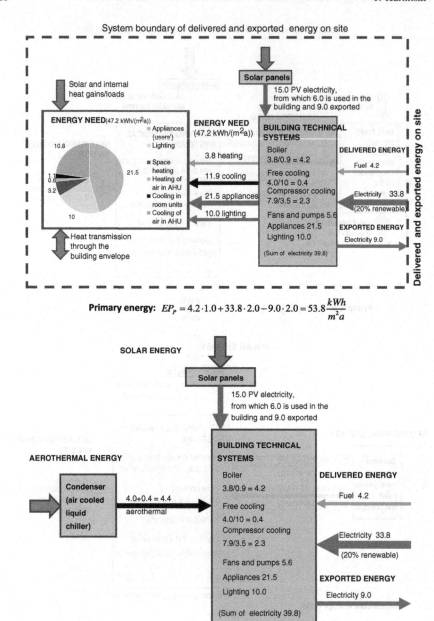

System boundary of delivered and exported energy on site

Primary energy: $EP_p = 4.2 \cdot 1.0 + 33.8 \cdot 2.0 - 9.0 \cdot 2.0 = 53.8 \dfrac{kWh}{m^2 a}$

Renewable energy ratio:

$$RER_p = \dfrac{15.0 + 4.4 + (2.2 - 2.0) \cdot 33.8}{15.0 + 4.4 + 4.2 \cdot 1.0 + 33.8 \cdot 2.2 - 9.0 \cdot 2.2} = \dfrac{26.2}{78.2} = 0.33$$

Fig. 10 Calculation example of the energy flows in nZEB office building

non-renewable primary energy factor of 2.5 is used. Therefore, the non-renewable primary energy factor of the grid mix is $0 * 0.2 + 2.5 * 0.8 = 2.0$, and the total primary energy factor is $1.0 * 0.2 + 2.5 * 0.8 = 2.2$. It is assumed that exported electricity compensates the grid mix.

From delivered and exported energy flows, non-renewable primary energy is calculated with the result of 53.8 kWh/(m^2 a). In the renewable energy ratio calculation, total primary energy factors are used, and the result is 33 %.

3.3 nZEB Office Building with Nearby Production

In this example, nearby wind farm electricity production is added to nZEB office building of Sect. 3.2. All other input data are from Sect. 3.2. A share of the wind farm production, corresponding to 20 kWh/(m^2 a) electricity, is allocated to the building. It is considered that 10 kWh/(m^2 a) of that production can be used in the building and 10 kWh/(m^2 a) is to be exported to the grid. These amounts can be calculated with hourly energy simulation of the building and hourly simulation of PV and wind electricity. The wind farm located nearby is connected to the grid.

The allocation of the wind farm production used in this example requires relevant legal and contractual framework so that the share of the wind farm capacity can be allocated to the building and the building owner will invest on new capacity of the wind farm correspondingly. In such a way, the new wind farm capacity means a real addition to the grid mix caused by and allocated to the building for a long term. This makes a major difference to Green Wash products that do not change the mix and do not add a new capacity.

With given assumptions, primary energy indicator and renewable energy ratio are calculated in Fig. 11.

3.4 On-Site CHP Production

The following input data are considered for the calculation example:

- Energy need for heating (space heating, supply air heating, and DHW) 35
- Electricity need 30
- Thermal renewable production on site 10
- Electric renewable production on site 9
- Combined heat and power on site:

  - Gas use 100
  - Produced heat 45
  - Thermal losses 20
  - Produced electricity 35
  - Auxiliary energy 0 (for simplicity).

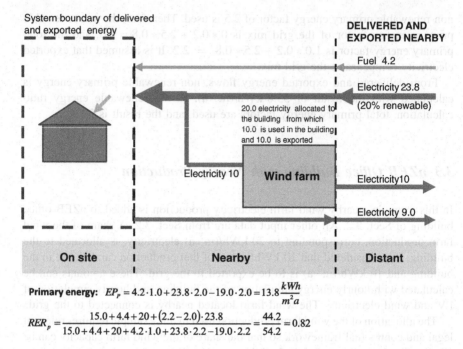

Primary energy: $EP_P = 4.2 \cdot 1.0 + 23.8 \cdot 2.0 - 19.0 \cdot 2.0 = 13.8 \dfrac{kWh}{m^2 a}$

$RER_P = \dfrac{15.0 + 4.4 + 20 + (2.2 - 2.0) \cdot 23.8}{15.0 + 4.4 + 20 + 4.2 \cdot 1.0 + 23.8 \cdot 2.2 - 19.0 \cdot 2.2} = \dfrac{44.2}{54.2} = 0.82$

Fig. 11 Calculation example of the energy flows in nZEB office building with nearby production

With this input data, the delivered gas becomes 100. It is assumed that from the produced electricity $35 + 9 = 44$, the exported electricity is 20. (This needs separate calculation and depends on the use and production data.) With exported electricity 20, the delivered electricity becomes $30 - (44 - 20) = 6$.

The non-renewable and total primary energy factors for delivered and exported energy are used as follows:

- Gas, total primary energy factor of 1.0 and non-renewable primary energy factor of 1.0;
- Delivered electricity, total primary energy factor 2.5, which corresponds to production from fossil fuel with 40 % efficiency ($1/0.4 = 2.5$);
- Exported electricity, total primary energy factor 2.5. It is considered that exported electricity compensates the grid mix.

Additionally, it is taken into account that there is the following renewable fuel share in the delivered energy:

- 10 % of delivered gas is renewable biogas with non-renewable primary energy factor 0 and total primary energy factor 1.0. The non-renewable primary energy factor of the delivered gas mix is $0 * 0.1 + 1.0 * 0.9 = 0.9$, and the total primary energy factor is $1.0 * 0.1 + 1.0 * 0.9 = 1.0$

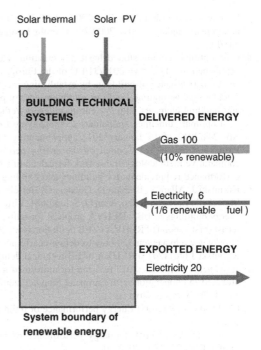

Primary energy:

$$EP_P = 100 \cdot 0.9 + 6 \cdot 2.083 - 20 \cdot 2.083 = 60.8 \frac{kWh}{m^2 a}$$

Renewable energy ratio:

$$RER_P = \frac{10 + 9 + (1.0 - 0.9) \cdot 100 + (2.5 - 2.083) \cdot 6}{10 + 9 + 100 \cdot 1.0 + 6 \cdot 2.5 - 20 \cdot 2.5} = \frac{31.5}{84.0} = 0.38$$

Fig. 12 Calculation example with on-site CHP production

- In the grid electricity production 1/6 of fuels are renewable. With 40 % production efficiency this results for non-renewable primary energy factor (1 - 1/6)/0.4 = 2.083.

Calculation of the non-renewable primary energy and renewable energy ratio (based on the total primary energy) with given input data is shown in Fig. 12.

References

1. EPBD (2010) Directive 2010/31/EU of the European parliament and of the council of 19 May 2010 on the energy performance of buildings (recast). http://ec.europa.eu/energy/efficiency/buildings/buildings_en.htm, http://eur-lex.europa.eu/JOHtml.do?uri=OJ:L:2010:153:SOM:EN:HTML
2. RES directive (2009) Directive 2009/28/EC of the European parliament and of the council. Apr 2009

3. Commission decision of 30 June 2009 establishing a template for National renewable energy action plans under directive 2009/28/EC of the European parliament and of the council 2009/548/EC

4. Cost optimal: commission delegated regulation (EU) No 244/2012 of 16 January 2012 supplementing. Directive 2010/31/EU of the European parliament and of the council on the energy performance of buildings by establishing a comparative methodology framework for calculating cost-optimal levels of minimum energy performance requirements for buildings and building elements. http://ec.europa.eu/energy/efficiency/buildings/buildings_en.htm

5. Cost optimal guidelines: guidelines accompanying Commission Delegated Regulation (EU) No. 244/2012 of 16 January 2012 supplementing Directive 2010/31/EU of the European Parliament and of the Council on the energy performance of buildings by establishing a comparative methodology framework for calculating cost-optimal levels of minimum energy performance requirements for buildings and building elements

6. Kurnitski J, Allard F, Braham D, Goeders G, Heiselberg P, Jagemar L, Kosonen R, Lebrun J, Mazzarella L, Railio J, Seppänen O, Schmidt M, Virta M (2011) How to define nearly net zero energy buildings nZEB—REHVA proposal for uniformed national implementation of EPBD recast (first version of REHVA nZEB definition from 2011). REHVA Eur HVAC J 48(3):6–12. http://www.rehva.eu/en/374.how-to-define-nearly-net-zero-energy-buildings-nzeb

7. Kurnitski J (ed) (2013) REHVA nZEB technical definition and system boundaries for nearly zero energy buildings, 2013 revision for uniformed national implementation of EPBD recast prepared in cooperation with European Standardization Organization CEN. REHVA Report No. 4, 2013. www.rehva.eu

8. EN 15603:2008 (2008) Energy performance of buildings—overall energy use and definition of energy ratings

9. prEN 15603:2013 (2013) Energy performance of buildings—overarching standard EPBD (draft), CEN/TC 371, 2012-10

10. EN 15251:2007 (2007) Indoor environment input parameters for design and assessment of energy performance of buildings addressing indoor air quality, thermal environment, lighting and acoustics

11. ISO 15927-4:2005 (2005) Hygrothermal performance of buildings: calculation and presentation of climatic data—Part 4: hourly data for assessing the annual energy use for heating and cooling

12. Jokisalo J, Kurnitski J (2007) Performance of EN ISO 13790 utilisation factor heat demand calculation method in a cold climate. Energ Buildings 39:236–247

13. Kurnitski J, Kalamees T, Tark T (2012) Early stage CAD-compliant energy performance assessment method. In: Seventh international cold climate HVAC conference, paper 24. Calgary, Canada, 12–14, Nov 2012

14. Marszal AJ, Heiselberg P, Bourrelle JS, Musall E, Voss K, Sartori I, Napolitano A (2011) Zero energy building: a review of definitions and calculation methodologies. Energ Buildings 43(4):971–979

15. Voss K, Sartori I, Lollini R (2012) Nearly-zero, net zero and plus energy buildings: how definitions and regulations influence solutions. REHVA Eur HVAC J 6(2012):23–27

16. Sartori I, Napolitano A, Voss K (2012) Net zero energy buildings: a consistent definition framework, energy and buildings, 48(2011):220–232. Available at: http://task40.iea-shc.org/publications

Present Energy Performance Requirements and nZEB Targets in Some Selected Countries

Jarek Kurnitski, Christian Feldmann, Per Heiselberg,
Livio Mazzarella, Igor Sartori, Karsten Voss and Åsa Wahlström

Abstract At the moment, there are already some official definitions of nZEBs available at least in Denmark, Estonia, and France, but most of the Member States intensively work with national definitions and plan for nZEBs. In the following, the situation in some selected countries is reported. At the end of the chapter, the issue of comparison of national requirements is discussed and comparison results of some countries based on the data of January 2013 are shown. Some harmonization in minimum energy performance requirements can be seen.

1 Denmark

The Danish Building Code (BR10) defines minimum energy performance requirements in terms of primary energy indicator for all new buildings [1]. It also includes two voluntary low-energy classes, class 2015 and class 2020, which reflects the expected future minimum energy performance requirements in 2015 and 2020, respectively.

The total primary energy use in the energy frame consists of heating, ventilation, cooling, domestic/service hot water, and lighting (except in residences). Tenants' or users' electricity is excluded. Heating (natural gas, oil or district heating) has a primary energy factor of 1, but factor of 0.8 and 0.6 can be used for district heating for buildings fulfilling class 2015 or class 2020, respectively. Electricity has a primary energy factor of 2.5, but for buildings fulfilling class 2020, a factor of 1.8 can be used. The floor area, A, used is the gross floor area measured outside the external walls. As a small country, there is only one climate zone.

J. Kurnitski (✉) · C. Feldmann · P. Heiselberg · L. Mazzarella · I. Sartori · K. Voss · Å. Wahlström
Tallinn University of Technology, Ehitajate tee 5, 19086, Tallinn, Estonia
e-mail: jarek.kurnitski@ttu.ee

J. Kurnitski (ed.), *Cost Optimal and Nearly Zero-Energy Buildings (nZEB)*,
Green Energy and Technology, DOI: 10.1007/978-1-4471-5610-9_3,
© Springer-Verlag London 2013

Table 1 Primary energy frames for new buildings in Denmark 2006, 2010, 2015, and 2020

Building code	Energy frame [kWh/(m^2 a)]			
	BR06	BR10	BR10—class 2015	BR10—class 2020
Residential	70 + 2200/A	52.5 + 1650/A	30 + 1000/A	20
Non-residential	95 + 2200/A	71.3 + 1650/A	41 + 1000/A	25

On-site renewable energy production is inside the system boundary and is subtracted from energy use, when calculating the delivered energy. In the energy performance calculation, the yearly production is subtracted from the calculated energy use, independently whether it is used directly in the building or exported to the grid for later use. (When exporting, the building owner only receives about 2/3 of the price of electricity compared to the price for buying. Therefore, it is economically beneficial to use electricity directly as much as possible and minimize export.)

In case of overheating, the energy use necessary to cool the building is included in the energy frame, even if no mechanical cooling system is available.

The building code also includes a number of specific requirements for each of the three energy classes to building components (e.g., heat loss through fabric, heat balance of windows, airtightness) and building systems (e.g., ventilation heat recovery efficiency, specific fan power, COP for heat pumps, efficiency of boilers). For energy class 2020, the building code also includes special requirements to the indoor environment (e.g., daylight availability, maximum overheating hours, maximum CO_2 level) (Table 1).

2 Estonia

Estonian regulation [2] defines minimum energy performance requirements in terms of the primary energy indicator for nearly zero-energy buildings, low-energy buildings, and for all new buildings as well as for major renovation. In minimum energy performance requirements, the results of cost-optimal analyses [3] are implemented, i.e., these mandatory requirements are tightened compared to previous regulation from 2007. Requirements for nZEB and low-energy buildings are not mandatory, but shall be followed, if nZEB or low-energy building (corresponding to *A* or *B* class of energy performance certificate) is constructed. For these requirements, the same format, i.e., nZEB building shall comply with the given limit value for primary energy indicator, is used as for minimum energy performance requirements. nZEB requirements are expected to come into force according to EPBD schedule.

Estonian energy performance requirements are based on primary energy, and the assessment boundaries follow the original REHVA definition from 2011. Energy frame is "all inclusive", i.e., heating, ventilation, cooling, domestic hot water, lighting, HVAC electricity, and users' appliances are all included. Electricity for appliances is defined by the standard use of the building types as well as operation times and occupancy profiles, ventilation rates, and the need for

Table 2 Estonian primary energy requirements (VV No. 68: 2012), which came into force since 9 Jan, 2013

	nZEB A kWh/(m² a)	Low energy B kWh/(m² a)	Min.req. new C (cost opt.) kWh/(m² a)	Min.req. maj.req. D (cost opt.) kWh/(m² a)
Detached houses	50	120	160	210
Apartment buildings	100	120	150	180
Office buildings	100	130	160	210

The requirements and corresponding energy certificate classes are shown in terms of primary energy for three building types out of nine

domestic hot water. For the lighting, tabulated or calculated values may be used. Requirements (and standard use) are given for nine building types, from which three are shown in Table 2.

Primary energy is calculated with non-renewable primary energy factors, which are for electricity 2.0, district heating 0.9, fossil fuels 1.0, and renewable fuels 0.75. On-site renewable energy production is inside the assessment boundary and is subtracted from energy use when calculating delivered and exported energy. The same primary energy factors are applied for delivered and exported energy.

For all buildings equipped with cooling, energy performance calculation/compliance assessment should be based on dynamic building simulation. Requirements are specified for simulation tools, which refer to relevant European, ISO, ASHRAE or CIBSE standards, IEA BESTEST or other equivalent generally accepted method. For residential buildings without cooling, monthly energy calculation methods may be also used. Exception is for detached houses, which have an alternative compliance assessment method based on tabulated specific heat loss values. This alternative method for houses may be used, if heat recovery ventilation with temperature ratio of at least 80 % and specific fan power no more than 2.0 W/(L/s) is used and the house has one of the listed heating systems. Tabulated specific heat loss values to be fulfilled depending on the heating system and values are given for ground source and air-to-water heat pumps, pellet boiler, district heating, and gas boiler.

In all buildings, dynamic temperature simulation in critical rooms is required in order to comply with summer temperature requirements (25 °C + 100 °C h in non-residential and 27 °C + 150 °C h in residential buildings during three summer months simulated with TRY). Only exception is for detached house; there the compliance may be alternatively shown with tabulated values for solar protection, window sizes, and window airing.

3 France

The new French regulation (RT2012) issued on October 26, 2010, and December 28, 2012, addresses nearly zero-energy building targets for residential buildings, office buildings, school buildings, kindergartens, hotels, retails, airport buildings, hospital buildings, restaurants, etc. RT 2012 building regulation came into force for both residential and non-residential buildings on January 1, 2013.

The total primary energy consumption is defined for heating, cooling, hot water production, lighting, ventilation, and any auxiliary systems used for these domains. It is given by an overall coefficient C_{ep} kWh/(m^2 a) using the net floor area of the building defined by the French building code.

The target maximum value of C_{ep}, C_{epmax} is fixed to 50 kWh/(m^2 a) with various correction coefficients depending on the type of building, the climatic zone, the altitude, the total area of the building, and the type of energy used.

Furthermore, in order to ensure a good quality of the design of the envelope, another constraint is required. A new parameter, B_{bio}, is added in order to check the "bioclimatic" quality of the design. B_{bio}, dimensionless parameter, assesses the ability of the building design to lower heat losses and air leakages of the envelope and optimize solar heat gains and natural lighting before the choice of any heating, air-conditioning, or ventilation system. B_{bio} is evaluated on the basis of a certain number of points. It has to be lower than **B_{bio} max** defined in the new regulation as a function of the location, altitude, type of building, etc.

Finally, the air tightness of the building is also imposed to a maximum value depending on the building type, and in summer, a limit for indoor summer temperature has to be checked if no cooling is used.

In order to prepare the 2020 objective of the "Grenelle de l'Environnement" policy, new labels will be in force in 2013 aiming 10 and 20 % energy consumption reduction more than the requirements of 2012 building regulation.

Consequently, "HPE" Label (for High-Energy Performance) will require 45 kWh/(m^2 a) as maximum annual energy consumption and "THPE" Label (Very High-Energy Performance) 40 (kWh/(m^2 a). In a first step, HPE and THPE Labels will be applied to residential, office, and school buildings.

Another label "BEPOS-Effinergie +" handled by EFFINERGIE ASSOCIA-TION is defined as an objective for 2020. Buildings will have to be "zero primary energy" on the basis of their consumption and production energy balance.

4 Germany

The current requirements (EnEV2009) for new buildings are calculated in relation to a so-called "reference building" with the identical geometry. For the reference building, standard envelope properties, such as U-values and standard installation engineering given in the EnEV, are applied within the calculation. The primary energy use of the planned building must be below or equal to the energy use of the reference building. Also, a limit value for the specific transmission heat loss is applied, and a minimum amount of heat from renewable sources should be reached (EEWärmeG). A primary energy conversion factor for electricity of 2.7 is being used. The procedure for non-residential buildings includes energy for space cooling and lighting into the building energy balance, whereas the focus for residential buildings are space heating, DHW, and ventilation only. A simplified energy calculation may be applied for residential buildings (DIN V 4108-6: 2003–06 and DIN V 4701-10: 2003–08), whereas all other buildings are calculated based on DIN V 18599:2007. The EnEV 2009 for the first time in Germany includes a method for crediting PV yields of on-site systems for the building energy balance (Sect. 5). Crediting is limited on a monthly basis upto the buildings monthly electricity use. Summer excess yield is taken as part of the grid and not accounted for in the building energy balance.

A special funding program of the Federal Building Ministry addresses a pilot market of net plus energy residential buildings. For this program, a modified calculation framework has been announced. The calculation is based on DIN V 18599, but credits the summer excess PV yield for the annual energy balance (seasonal shift) and includes household appliances on a fixed basis in the overall energy balance. Primary energy factors for electricity are modified to 2.4 for electricity delivered and 2.8 for exported on-site generation (asymmetric weighting factors). The energy balance has to be positive (weighed exported > weighed delivered) for end energy as well as primary energy.

Official definitions concerning the public subsidies for (residential) low-energy buildings are the subject of the programs run by the (state-owned) Kreditanstalt für Wiederaufbau Frankfurt (KfW). These programs are mainly fed by public sources. The current requirements are KfW 70, KfW 55, and KfW 40. The primary energy demand of these buildings has to be 70, 55, and 40 % of the reference building. In addition, there is also a subsidy program for "Passiv-Häuser", which is defined in accordance with the Passiv-Haus-Institute as "KfW-40-buildings with an annual space heating demand lower than 15 kWh/m^2".

Discussions on the next version of the EnEV are going on but have not been finalized (status 1/2013). Focus of the discussion is the cost-effectiveness of further energy reduction measures on the one hand and the political aim of "climate neutrality" for the building sector in future on the other hand.

5 Italy

The building energy use is regulated at the regional level by each local Government; this means, existing 20 regions, that 20 different regulations and codes are in principle possible. Actually, there are nine regions (Valle d'Aosta, Piemonte, Lombardia, Friuli Venezia Giulia, Liguria, Emilia Romagna, Toscana, Puglia, and Sicilia) and one autonomous province (Bolzano), which have had directly employed by their own the old EPBD directive and have slightly different methodologies to calculate the building energy performance. For the other nine regions (Veneto, Marche, Umbria, Lazio, Molise, Abruzzo, Campania, Basilicata, and Calabria) and the remaining autonomous province (Trento), the substitutive national law is in force instead. Today, there is the intention to harmonize the regional legislations through the country when employing the EPBD recast. For

Table 3 Heating primary energy performance indicator (primary energy use by useful floor area) for residential buildings

Shape ratio S/V	Residential buildings, kWh/(m² a)								
				Climatic zone					
	A	B		C		D		E		F
	Up to 600 DD	From 601 DD	To 900 DD	From 901 DD	To 1,400 DD	From 1,401 DD	To 2,100 DD	From 2,101 DD	To 3,000 DD	Over 3,000 DD
<0.2	8.5	8.5	12.8	12.8	21.3	21.3	34	34	46.8	46.8
>0.9	36	36	48	48	68	68	88	88	116	116

Table 4 Heating primary energy performance indicator (primary energy use by useful floor area) for non-residential buildings

Shape ratio S/V	non-residential buildings, kWh/(m³ a)								
				climatic zone					
	A	B		C		D		E		F
	Up to 600 DD	From 601 DD	To 900 DD	From 901 DD	To 1,400 DD	From 1,401 DD	To 2,100 DD	From 2,101 DD	To 3,000 DD	Over 3,000 DD
<0.2	2.0	2.0	3.6	3.6	6	6	9.6	9.6	12.7	12.7
>0.9	8.2	8.2	12.8	12.8	17.3	17.3	22.5	22.5	31	31

this reason, the national government is working today to provide new country-harmonized legislation and procedure defining country-wide cost-optimal energy performance limits.

To give an idea of what will be the future cost-optimal energy performance limits, the new buildings (nZEBs) have to comply with the actual limits reported in the national law for heating only, which are reported (Tables 3, 4).

In principle, the requirement for the nZEB will be less or equal to these actual values. Up to now, lighting and Tenants' or users' electricity is excluded. Heating (natural gas, oil) has a primary energy factor of 1; the factor used for district heating for buildings has to be declared by the district heating companies. Electricity has a primary energy factor of 2.18. Cooling performance in not evaluated in terms of energy use but only of energy need, giving some limiting values for the required thermal energy by useful floor area.

6 Norway

A low-energy commission (set up by the Ministry of Petroleum and Energy) delivered a number of suggestions for increased energy efficiency of all sectors in Norway in 2009, including suggestions of future net energy frame values for new buildings as well as for major renovations [4]. In 2012, two ministerial recommendations to the Parliament from the Ministry of Environment and the Ministry of Local Government and Regional Development, respectively, announced the adoption of the passive house standard for all new buildings from 2015 and the nearly zero-energy standard from 2020 [5].

The Norwegian Building Code, TEK, is proposed to be sharpened every fifth year. TEK07, published in 2007, was the first in Norway with an energy performance approach. The net energy (energy needs) in the energy frame consists of heating, ventilation, cooling, domestic/service hot water, as well as lighting and tenants' or users' electricity. The net energy includes cooling supplied to air-cooling coils or fan coils in the rooms. The building code has already been updated in 2010 [6] and will therefore be sharpened further in 2015 to implement the passive house standard in Norway, which is defined by the norms NS 3700 [7] for residential buildings and NS 3701 [8] for non-residential buildings. The same norms contain the definition of "low-energy building", based on the same method but with less stringent parameters than the passive house standard. The low-energy building standard may be adopted as the target for major renovations (not agreed yet). Today, major renovations have to comply with TEK10, as long as technical or architectural conditions do not make these non-economic. The definition of nZEB to be adopted as a standard for new buildings from 2020 is still under development.

The floor area used is the heated floor area measured inside the external walls (including internal partitions). Norway has a number of climate zones. The values given below are valid for the "standard" climate zone around Oslo, which is in the southeastern part of the country. The annual energy use of the proposed building is first modeled for the actual climate zone and then for the "standard" climate zone. The results for the standard climate zone must fulfill the energy frame. The current energy frames are specified for one-family houses, multi-family houses, and eleven types of non-residential buildings (office given as a reference) (Table 5).

Table 5 Proposed future net energy frames for new buildings in Norway

Building code	Energy frame [kWh/m^2 y]					
	TEK07	TEK10	TEK15—Passive house	TEK20	TEK25	TEK30
Residential (detached house)	135	130	80 (Heating: 15, cooling: 0, DHW: 30)	Nearly ZEB	Intermediate	Net ZEB
Residential (apartment block)	120	115				
Non-residential (office)	165	150	75[a] (Heating: 20, cooling: 10, DHW: 5)			

[a] The low value is largely due to improvement in electrical appliances and adoption of demand-controlled lighting and ventilation, on top of envelope improvements. Furthermore, the low amount of hot water required in offices makes the total energy need lower than that for residential units

7 Sweden

The most recent edition of the building regulations published by the National Board of Housing, Building and Planning dates from 2006. It sets out the requirements for energy performance of buildings. Further restrictions for electrically heated buildings were published in 2009, and further restrictions for buildings heated with other than electrical heating were published in 2012 [9]. The regulations are supposed to be revised every third year, and a specific control point on how to define nZEB is planned for 2015. Large work is now going on both by the Swedish Energy Agency and the National Board of Housing, Building and Planning in collecting knowledge and experiences for that control point.

The requirements specify not only maximum permitted delivered energy use per square meter, but also the permitted installed electric power for heating and a mean coefficient of thermal transmittance of the building envelope. In addition, the new building code specifies that energy performance must be verified by measurements within 24 months of completion of the building.

The requirements are described in terms of delivered energy use (kWh/m^2 A_{temp}) and are shown in Table 6. A_{temp}, the temperate area, is defined as the area on the inside of the building envelope, on all floors, that is supposed to be heated to more than 10 °C. The area of interior walls, openings for stairs, shafts, and similar are included. The area of the garage is not included.

The delivered energy use is defined as the energy that needs to be delivered to the building (delivered energy is often called "purchased energy" in Sweden), at normal use and during a normal year, for heating, comfort cooling, hot tap water, and electricity for the operation of the building. It can be reduced by energy contributions from solar cells and solar collectors installed on the building. Electricity for domestic purposes in residential buildings or business activities in premises is not included.

The requirements differ depending on the following: in which climate zone the building is placed (Sweden is divided into three climate zones, shown in Table 6), whether the building has an occupant activity of living (dwellings) or business activities (premises), and whether the building is heated by electricity or in another way. About 80 % of the population lives in southern climate zone, and less than 10 % lives in the northern climate zone.

Electrically heated buildings are buildings in which the installed electric power for heating is greater than 10 W/m^2. Installed power is the total electric power that can be delivered by the electrical heating devices that are needed to maintain the intended indoor climate, domestic hot water production, and ventilation when the maximum power needs of the building prevail, that is, during the design outdoor winter temperature. This means that requirements for buildings with electric heating apply also for most of heat pump systems.

The new regulation that applies from January 1, 2013, also has stricter rules for making changes in buildings. These rules state that the same quality requirements

Table 6 Requirements for delivered energy use in the national building code that applies from the January 1, 2013 (kWh/m² At_emp)

	Annual delivered energy use for heating, comfort cooling, domestic hot water provision, and other shared services in the building (kWh/m² a)		
	Climate zone		
	1 (North Sweden)	2 (Middle Sweden)	3 (South Sweden)
Residential buildings with heating systems other than electric heating	130	110	90
Residential buildings with electric heating	95	75	55
Commercial and similar premises with heating systems other than electric heating	$120 + 110 \times (q - 0.35)$	$110 + 90 \times (q - 0.35)$	$80 + 70 \times (q - 0.35)$
Commercial and similar premises with electric heating	$95 + 65 \times (q - 0.35)$	$75 + 55 \times (q - 0.35)$	$55 + 45 \times (q - 0.35)$

Climate Zone 1

Climate Zone 2

Climate Zone 3

[1] q is the average specific outdoor air ventilation flow rate during the heating season (l/(s, m²)) and is an addition that must be included when the outdoor air flow exceeds 0.35 l/(s, m²) in order to maintain required hygienic air quality in temperature-controlled areas. Its maximum permissible value is 1.00 l/(s, m²)

that apply to the construction of a new building will apply to changes in buildings. This means, in actuality, that energy performance requirements for new buildings will also apply to extensive renovations of an existing building.

8 How to Compare National Energy Performance Requirements?

It is essential to compare energy performance requirements used in different countries. However, in most cases, it cannot be directly seen which country has more stringent requirements. The comparison is challenging because of differences in energy calculation frames, input data, and calculation rules (and calculation tools) as well as in climate data. The use of primary energy as energy performance indicator, required by EPBD, is not yet implemented in all countries. Some national energy performance frameworks still use the delivered energy (i.e., energy purchased to building without primary energy factors), and in some countries, heat sources (heat pumps, etc.) are not taken into account. National energy frames may also include different energy flows taken into account and household electricity, and lighting is sometimes taken into account and sometimes not in residential buildings. In non-residential buildings, lighting is typically considered, but appliances (plug loads) are not taken into account in all countries. For this reason, the quantitative comparison needs calculations with assumptions, in order to shift national energy performance requirements to delivered energy, which can be most easily compared in all countries. Also, a degree-day correction of space heating is needed when comparing countries with central European and Nordic climate.

In the following, the maximum allowed delivered energy is calculated from national requirements of five countries for residential houses, apartment buildings, and office buildings. The differences in the input data of lighting, appliances, and domestic hot water (DHW) were taken into account. Energy use of DHW (national variation between 15 and 35 kWh/(m^2 a)) was normalized to 25 kWh/(m^2 a) in all countries by increasing or decreasing the maximum allowed delivered energy values. After that, in residential buildings, the maximum allowed delivered energy without household electricity and lighting was calculated, because these components are not included in energy frames of some countries. If included, the values of these components were reduced. Therefore, the maximum allowed delivered energy in the comparison includes space and ventilation heating, domestic hot water heating (with normalized DHW need), cooling, and HVAC electricity for running fans and pumps, etc. For office buildings, the lighting was also included, but user appliances (plug loads) were not included.

The national requirements (data from January 2013) needed in the comparison are listed below in condensed format for each country.

Denmark BR2010:

- Primary energy $52.5 + 1650/A_{\text{gross}}$ kWh/(m^2 a) for residential and $71.3 + 1650/A_{\text{gross}}$ kWh/(m^2 a) for non-residential (the values do not include the household electricity and lighting in residential and appliances in non-residential)
- Energy need of DHW $250\,\text{L}/\left(m^2_{\text{gross}}a\right)$ $14.5\,\text{kWh}/\left(m^2_{\text{gross}}a\right)$ in residential
- Primary energy factor 2.5 for the electricity and 1.0 for oil, gas, and district heat
- Primary energy values were shifted to the net area, for houses $A_{\text{gross}}/A_{\text{net}} = 1.17$, and for other buildings, $A_{\text{gross}}/A_{\text{net}} = 1.05$ was used

Sweden 2012:

- Delivered energy 90 kWh/(m^2 a) in South region in residential buildings
- 55 kWh/(m^2 a) in the case of electrical (and heat pump) heating in residential buildings
- $80 + 70(q_{\text{avg}} - 0.35)$ kWh/(m^2 a) where q_{avg} is annual average outdoor air flow rate in $L/(s\ m^2)$ ($q_{\text{avg}} = 0.49$ was used) for non-residential buildings, and +15 kWh/(m^2 a) was considered for lighting
- The values do not include the household electricity and lighting
- Energy need of DHW 20 in houses and 25 kWh/(m^2 a) in apartments

Norway TEK 2010:

- Energy use without generation is $120 + 1600/A_{\text{heated}}$ kWh/(m^2 a) in detached houses, 115 in apartment buildings, and 150 kWh/(m^2) in offices
- The values include household electricity and lighting of 28.9 from which heat gain is 21.9 kWh/(m^2 a) in residential, and appliances of 34 kWh/(m^2 a) in offices, heat pumps are not taken into account
- Energy Need of DHW 29.8 kWh/(m^2 a) in Residential

Estonia 2012:

- Primary energy is 160 in houses, 150 in apartments, and 160 kWh/(m^2 a) in offices, all of which include household electricity and lighting; without household electricity and lighting, primary energy is $160 - 50 = 110$ in houses and $150 - 59 = 91$ in apartments and without appliances $160 - 38 = 122$ kWh/(m^2 a) in offices
- Primary energy factor 2.0 for the electricity, 1.0 for oil and gas, and 0.9 for district heat
- Energy need of DHW 25 kWh/(m^2 a) in houses and 30 in apartments

Finland D3 2012:

- Primary energy is $372 - 1.4 \times A_{\text{net}}$ in houses, 130 in apartments, and 170 kWh/(m^2 a) in offices, all of which include household electricity and lighting; without household electricity and lighting, primary energy is $162 - 39 = 123$ in 150 m^2 house and $130 - 52 = 78$ in apartments and without appliances $170 - 38 = 132$ kWh/(m^2 a) in offices

- Primary energy factor 1.7 for the electricity, 1.0 for oil and gas, and 0.7 for district heat
- Energy need of DHW 35 kWh/(m^2 a) in residential

Because the requirements may depend on the size of building for detached houses, a 150 m^2 house was considered. In the countries where the requirements included household electricity and lighting, the values of these were reduced from the delivered energy requirement (or from the delivered energy value corresponding to the primary energy requirement). Energy use of DHW was normalized to 25 kWh/(m^2 a) in residential buildings with corresponding corrections to the delivered energy requirement. In the countries where the energy use of DHW was smaller than 25 kWh/(m^2 a), the corresponding difference in the delivered energy was reduced from the delivered energy requirement, and in countries with the larger value, the difference was added. System losses of DHW were neglected in this normalization.

To take into account the electricity use of HVAC systems with the primary energy factor (if exists), the electricity use for fans of ventilation and for circulation pumps of water-based heating was specified. To enable degree-day correction for space heating, the normalized value of DHW was reduced for heating energy use. In the calculations, the following assumptions were used:

- Energy use for domestic hot water heating of 25 kWh/(m^2 a) is for houses and apartment buildings and 6 kWh/(m^2 a) for office buildings
- Electricity use of 5 kWh/(m^2 a) for fans of ventilation and 3 kWh/(m^2 a) for circulation pumps of water-based heating (0 kWh/(m^2 a) for electrical heating) in houses and 7 kWh/(m^2 a) for fans and 2 kWh/(m^2 a) for circulation pump in apartment and office buildings
- Electricity use according to national values for appliances and lighting in houses and apartment buildings and for appliances in office buildings
- Remaining space heating energy was corrected with degree-day correction in relation to Copenhagen degree-day value

Figure 1 shows the maximum allowed delivered energy for houses without household electricity and lighting (i.e., delivered energy to heating, hot water, and ventilation systems) in each country. The values are degree-day corrected, because the space heating differs between the coldest (Helsinki) and warmest (Copenhagen) by a factor of 1.4. Degree-day for 17 °C base temperature were calculated from ASHRAE 2001 climate data for each capital city: 3259 °C d Copenhagen, 3894 °C d Oslo, 3963 °C d Stockholm, 4240 °C d EstoniaTRY, 4422 °C d Helsinki.

Calculation principle of values in Fig. 1 can be explained with the following examples. The value of the Denmark/electrical is calculated from the primary energy requirement per net area, that is, $(52.5 + 1,650/150) \times 1.17 = 74.3$, where $1.17 = A_{gross}/A_{net}$. To normalize DHW need from Danish 17.0 to 25, the difference has to be added. The delivered electrical energy w/o households

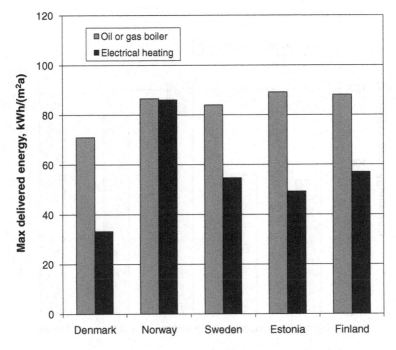

Fig. 1 Maximum allowed delivered energy for heating, hot water (with normalized use), and ventilation systems for 150 m^2 house in each country; degree-day corrected data

becomes $(74.3 + 8.0)/2.5 = 33.2$, where 2.5 is primary energy factor (there is no degree-day correction in Danish case).

The value of Norway/oil or gas is calculated from the requirement $(120 + 1,600)/150 = 130.7$. Then, household electricity and lighting (28.9) is reduced, and DHW (29.8) is normalized: $130.7 - 28.9 - 4.8 = 97.0$. Next step is to calculate space heating energy for the degree-day correction. Fans, pumps, and normalized DHW are to be subtracted: $97.0 - 5 - 3 - 25 = 64.0$. The degree-day correction is $64.0 \times 3,894/3,259 = 53.5$. From corrected space heating energy, the delivered energy is calculated by adding fans, pumps, and DHW: $53.5 + 5 + 3 + 25 = 86.5$.

To compare requirements for houses with ground source heat pumps, the net energy demand for space, ventilation, and hot water heating can be compared. This was calculated with the seasonal energy efficiency ratio of 3.5 for space and ventilation heating and 2.5 for domestic hot water heating, Fig. 2.

Similar to Fig. 1, the maximum allowed delivered energy for apartment and office buildings is shown in Fig. 3. District heat is considered as heat source, but the results will be similar with a gas boiler. Appliances and lighting are not included in apartment buildings, but in the office buildings, lighting is included in the delivered energy.

It has to be noted that the comparison given has some limitations. The differences in national input data regarding lighting, appliances, and hot water were

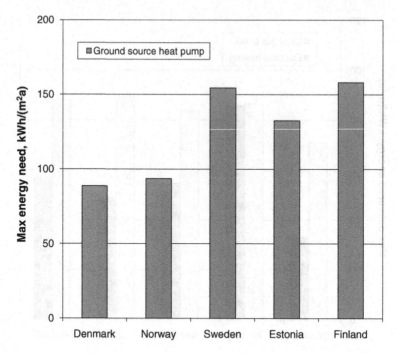

Fig. 2 Maximum allowed net energy need for space, ventilation heating, and hot water heating (normalized use) with heat pumps

taken into account. DHW was normalized, but the lighting and appliances were just reduced from delivered energy. This reduction will give some advantage for the countries with high energy use values for lighting and appliances as high internal heat gains reduce the need for heating. In principle, it is possible to apply adjustment for heating energy (the gain utilization factor has to be estimated), but this was not done in the comparison, because the gain utilization factor assessment would have been too speculative without running simulations of reference build-ings. Other factors not possible to take into account in this type of comparison are differences in national ventilation rates, internal heat gains from occupants, heating set points, as well as differences in national calculation methods/tools, which may use different approaches and parameters.

A more detailed simulation-based comparison with reference buildings, detailed national input data, real weather data, and national calculation tools is reported in [10]. According to these more detailed results, the simplified method reported here underestimated the max delivered energy in Denmark by about 10 %, but Danish requirements were most strict according to both methods (simplified method was not able to take into account input data differences in ventilation rate and occupancy and 1 °C lower heating set point in Denmark). As a rule, the simplified method did not change the order of countries in the compar-ison. Only exception was for the office building, for which the positions of

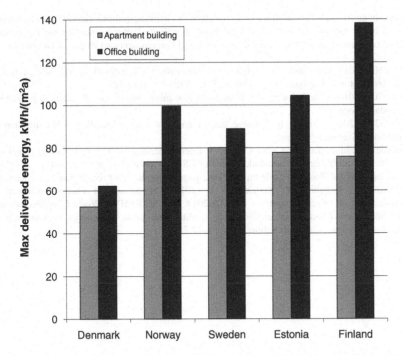

Fig. 3 Maximum allowed delivered energy for heating, hot water (with normalized use), and ventilation systems in apartment buildings and office buildings (lighting included) with district heating

Norwegian and Swedish buildings were changed compared to simplified method, i.e., the requirements were slightly more tight in Norway relative to Sweden when compared with the simulation-based method.

References

1. Denmark Danish Building Code BR2010, http://www.ebst.dk/file/104440/bygningsreglement. pdf (Older Building Codes, http://www.ebst.dk/br08.dk) (in Danish)
2. Estonia VV No 268 (2012) Estonian Government ordinance No. 68, 2012, Minimum requirements for energy performance of buildings, Vabariigi Valitsuse määrus Nr. 68 (30.08.2012). Energiatõhususe miinimumnõuded. *Riigi Teataja*, I, 05.09.2012, 4. https:// www.riigiteataja.ee/akt/105092012004 (in Estonian)
3. Kurnitski J, Saari A, Kalamees T, Vuolle M, Niemelä J, Tark T (2011) Cost optimal and nearly zero (nZEB) energy performance calculations for residential buildings with REHVA definition for nZEB national implementation. Energy and Buildings 43(11):3279–3288
4. Norway: Lavenergiutvalget—Energieffektivisering (2009) The low energy commission— energy efficiency. Ministry of Petroleum and Energy. http://www.regjeringen.no/upload/ OED/Rapporter/OED_Energieffektivisering_Lavopp.pdf (in Norwegian)

5. Gode bygg for eit bedre samfunn—Ein framtidsretta bygningspolitikk (2012) Good buildings for a better society—a forward-looking building policy. Announcement to the Parliament 28—Recommendations from the Ministry of Local Government and Regional Development, 15/06/2012
6. TEK10 (2010) Regulation on technical requirements for construction. Ministry of Local Government and Regional Development, FOR 2010-03-26 nr 489
7. NS 3700 (2010) Criteria for passive houses and low energy buildings—Residential buildings, Standard Norway
8. NS 3701 (2012) Criteria for passive houses and low energy buildings—Non-residential buildings, Standard Norway
9. Sweden Building Energy Code, Chapter 9 Energy Management. http://www.boverket.se/ Bygga–forvalta/Bygg–och-konstruktionsregler-ESK/Boverkets-byggregler/ (The English translation of the Building Code on the home page is not the most recent one!) (in Swedish)
10. Kurnitski J, Grönlund V, Reinikainen E (2013) Comparison of energy performance requirements in selected countries. CLIMA 2013. In: 11th REHVA World Congress and the 8th international conference on indoor air quality, ventilation and energy conservation in buildings, Prague, Czech Republic, June 16–19 2013

Cost Optimal Energy Performance

Jarek Kurnitski

Abstract EPBD recast requires Member States (MS) to ensure that minimum energy performance requirements of buildings are set with a view to achieving cost optimal levels using a comparative methodology framework established by the Commission [1]. Cost optimal performance level means the energy performance in terms of primary energy leading to minimum life cycle cost. MS had to provide first cost optimal calculations to evaluate the cost optimality of current minimum requirements due March 2013. After that, MS need to revise calculations and to submit reports to the Commission at regular intervals, which shall not be longer than 5 years. Cost optimal methodology is intended for the minimum energy performance requirements, but as well defined methodology, this can be used also in any construction project in order to find most cost optimal solutions. In the following, a systematic and robust procedure to determine cost optimal energy performance solutions is discussed.

1 General Methodology

For the systematic and robust cost optimal energy performance calculation procedure the following seven steps can be identified:

1. selection of the reference building/buildings
2. definition of construction concepts based on building envelope optimization for fixed specific heat loss levels [from business as usual (BAU) construction to highly insulated building envelope in four steps]
3. specification of building technical systems
4. energy calculations for specified construction concepts

J. Kurnitski (✉)
Tallinn University of Technology, Ehitajate tee 5, 19086, Tallinn, Estonia
e-mail: jarek.kurnitski@ttu.ee

J. Kurnitski (ed.), *Cost Optimal and Nearly Zero-Energy Buildings (nZEB)*,   47
Green Energy and Technology, DOI: 10.1007/978-1-4471-5610-9_4,
© Springer-Verlag London 2013

5. post processing of energy results to calculate delivered, exported and primary energy
6. economic calculations for construction cost and net present value (NPV) of operating cost
7. sensitivity analyses (discount rate, escalation of energy prices and other parameters)

All this steps are independent and they did not lead to iterative approach or optimization algorithm, which indeed can improve accuracy in more detailed analyses. Cost optimal calculation to obtain the minimum NPV can just done by straightforward calculation of steps 2–6 for all specified cases (according to steps 2 and 3). If specified cases will not show the minimum of the NPV, additional cases are to be specified to obtain the minimum.

Cost optimal primary energy use is determined by the solutions leading to minimum NPV of 30 years period for residential buildings and 20 years period for non-residential buildings according to the Cost optimal regulation [2, 3]. Reference buildings are needed for calculations. For new buildings, one representative reference building is enough, however it may provide valuable information if in the sensitivity analyses another reference building will be used. Construction concepts to be studied have to represent building envelopes from BAU construction to highly insulated building envelope. With building envelope optimization only four construction concepts are enough to change insulation thickness mainly with 5-cm step and with 10-cm step for thicker insulations. Heat recovery efficiency is the feature belonging to the construction concept, because of the gain utilization in energy calculations. To keep calculations simple, fixed heat recovery efficiency is to be used for each construction concept. All relevant heating (and cooling) systems can be calculated with reasonable effort, if the same distribution and emission systems will be used for all cases simplifying cost calculations and to ensuring equal comfort level.

General nZEB energy calculation framework described in Chap. 2 can be used for cost optimal energy performance calculations.

2 Input Data Selection Principles

Step 1. *The reference building*

It can be recommended that architects will select the reference building as a typical representative building of new construction. Single family building, multi family building and office building (one for new built and two for existing) are required by Cost optimal regulation. An example is calculated with Estonian detached reference house with heated net floor area of 171 m^2, Fig. 1.

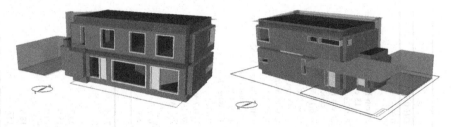

Fig. 1 Energy simulation model of the reference detached house. Perspective view from South to East in the *left* and from North to West in the *right*

Step 2. *Definition of construction concepts*

Proper definition of the construction concepts (=building envelope + heat recovery) is the cornerstone of the method. Careful selection of construction works allows to reduce calculation effort drastically. In the example, four construction concepts (Table 1) were specified based on the specific heat loss coefficient. The specific heat loss coefficient includes transmission and infiltration losses through the building envelope and is calculated per heated net floor area:

$$\frac{H}{A_{\text{floor}}} = \frac{\sum U_i \cdot A_i + \sum \Psi_j \cdot l_j + \sum \chi_p \cdot n_p + \rho_a \cdot c_a \cdot \dot{V}_i}{A_{\text{floor}}} \tag{1}$$

where:

H Heat loss coefficient, W/K;
A_{floor} Heated net floor area, m^2;
U_i Thermal transmittance of envelope part $_i$, W/(m^2 K);
A_i Area of envelope part $_i$, m^2;
Ψ_i Thermal conductance of linear thermal bridge $_i$, W/(mK);
l_i Length of linear thermal bridge $_i$, m;
χ_p Thermal conductance of point thermal bridge $_p$, W/K;
n_p Number of point thermal bridge $_p$, -;
ρ_a Density of air, kg/m^3;
c_a Specific heat capacity of air, J/(kg K);
$\dot{V}_i$ Infiltration rate, m^3/s:

Infiltration rate can be calculated with national regulation, here the Estonian equation [4] was used:

$$\dot{V}_i = \frac{q_{50} \cdot A_{\text{env}}}{3{,}600 \cdot x} \tag{2}$$

where:

q_{50} Air leakage rate of building envelope, m^3/(h m^2);
A_{env} Area of building envelope (including the bottom floor), m^2;

Table 1 Construction concepts and simulated net energy needs of the reference detached house of 171 m²

	Construction concepts			
	DH 0.42 "Nearly zero"	DH 0.58	DH 0.76	DH 0.96 "BAU"
Specific heat loss coefficient H/A, W/m² K	0.42	0.58	0.76	0.96
External wall 170 m²	20 cm LWA block, plaster + 35 cm EPS-insulation U 0.1 W/m² K	20 cm LWA block, plaster + 25 cm EPS-insulation U 0.14 W/m² K	20 cm LWA block, plaster + 20 cm EPS-insulation U 0.17 W/m² K	20 cm LWA block, plaster + 15 cm EPS-insulation U 0.23 W/m² K
Roof 93 m²	80 cm min.wool insulation, concrete slab U 0.06 W/m² K	50 cm min.wool insulation, concrete slab U 0.09 W/m² K	32 cm min.wool insulation, concrete slab U 0.14 W/m² K	25 cm min.wool insulation, concrete slab U 0.18 W/m² K
Ground floor 93 m²	Concrete slab on ground, 70 cm EPS U 0.06 W/m² K	Concrete slab on ground, 45 cm EPS U 0.09 W/m² K	Concrete slab on ground, 25 cm EPS U 0.14 W/m² K	Concrete slab on ground, 18 cm EPS U 0.18 W/m² K
Leakage rate q_{50}, m³/(h m²)	0.6	1.0	1.5	3.0
Windows 48 m² U-value glazing/frame/total	4 mm–16 mm Ar-SN 4 mm–16 mm Ar-SN4 mm Insulated frame 0.6/0.7 W/m² K 0.7 W/m² K	4 mm–16 mm Ar-4 mm–16 mm Ar-SN 4 mm Insulated frame 0.8/0.8 W/m² K 0.8 W/m² K	4 mm–16 mm –4 mm–16 mm Ar-SN 4 mm 1.0/1.3 W/m² K 1.1 W/m² K	4 mm–16 mm Ar-SN 4 mm Common frame 1.1/1.4 W/m² K 1.2 W/m² K
g-value	0.46	0.5	0.55	0.63
Ext. door, 6 m²	U 0.7 W/m² K	U 0.7 W/m² K	U 0.7 W/m² K	U 0.7 W/m² K
Ventilation rate l/s, specific fan power SFP, temperature efficiency AHU HR	80 l/s, SFP 1.5 kW/(m³/s), AHU HR 85 %	80 l/s, SFP 1.7 kW/(m³/s), AHU HR 80 %	80 l/s, SFP 2.0 kW/(m³/s), AHU HR 80 %	80 l/s, SFP 2.0 kW/(m³/s), AHU HR 80 %
Heating capacity, kW	5	6	8	9
Cooling capacity, kW	5	5	5	8
Energy need kWh/(m² a)				
Space heating	22.2	36.8	55.1	71.5
Supply air heating in AHU	4.1	5.7	5.7	5.7
Domestic hot water	29.3	29.3	29.3	29.3
Cooling	13.6	11.1	9.2	15.0
Fans and pumps	7.9	8.8	10.0	10.0
Lighting	7.3	7.3	7.3	7.3
Appliances	18.8	18.8	18.8	18.8
Total energy need	103.2	117.8	135.5	157.7

x Factor taking into account the height of the building: 35 for single-storey houses, 24 for 2-storey buildings, 20 for buildings for 3...4-storeys and 15 for ≥ 5 storeys;

DH 0.42 construction concept represents the best practice technology of highly insulated building envelope which may be associated with nearly zero energy buildings. DH 0.96 represents BAU construction. Building envelope has to be optimized for each specific heat loss value, so that the most cost effective combination of insulation levels for windows, external walls, slab on ground and roof will used to achieve the given specific heat loss value. This means that one has to select a proper window and external wall insulation combination, to achieve the given specific heat loss value at the lowest possible construction cost. This is a basic construction cost calculation exercise, the professionals are doing daily. If this is followed, one will need to calculate net energy needs only once (four simulations in this case).

Step 3. *Specification of building technical systems*

All cases were equipped with effective heat recovery (as in a cold climate) and were calculated with almost all possible heating systems. For each construction concept, the following heating systems were considered with appropriate sizing:

- ground source heat pump
- air to water heat pump
- district heating
- direct resistance electrical heating
- condensing gas boiler
- condensing oil boiler
- pellet boiler

Sizing data of the systems is shown in Table 1 and performance data in Table 2. Because of the cold climate and dominating heating need, only one basic compressor cooling solution was used for all cases. Highly insulated DH 0.42 and DH 0.58 cases were calculated both with and without solar collectors of 6 m^2, providing an half of domestic hot water net energy need. Other cases were calculated without solar collectors. For nZEB, 5 kW solar PV installation was additionally used.

In principle, the number of technical systems to be studied can be high, because of the fast post processing of energy calculation results. All relevant technical systems could be relatively easily calculated (resulting mainly as the effort for cost calculations) to be sure that the combination leading to minimum NPV will not missed due to limited systems specification.

Step 4. *Energy simulations for specified construction concepts*

All relevant energy calculation tools can be used, however the validated dynamic tools can be recommended. Such tools are contrasted in [5]. For the example, energy simulations were conducted with dynamic simulation tool

Table 2 System efficiencies for delivered energy calculation

Heat source (under floor heating)	Generation and distribution combined efficiency, -	
	Space heating/cooling	Domestic hot water
Gas/oil condensing boiler	0.86	0.83
Pellet boiler	0.77	0.77
Air to water heat pump (electricity)	1.98	1.62
Electrical heating	0.90	0.90
Ground source heat pump (electricity)	3.15	2.43
District heating	0.90	0.90
Cooling (electricity)	3.0	

IDA-ICE [6] for specified four construction concepts. Simulated net energy needs are shown in Table 1.

Step 5. *Post processing of the simulation results to calculate delivered, exported and primary energy*

Delivered energy can be easily calculated with post processing from simulated net energy needs. Net energy needs are to be divided with relevant system efficiencies. System efficiency values used in the example (combined efficiency of the generation, distribution and emission) are shown in Table 2. To calculate the combined efficiency, under floor heating distribution was considered with average distribution and emission efficiency of 0.9 according to Estonian regulation.

To calculate primary energy, exported energy has to be reduced from delivered energy. National primary energy factors are to be used, the example used Estonian ones:

- fossil fuels 1.0
- electricity 2.0
- district heating 0.9
- renewable fuels 0.75

Step 6. *Economic calculations: construction cost and NPV calculations*

Economic calculations include construction cost calculations and discounted energy cost calculation for 30 years. To save calculation effort, construction cost is accepted to calculate not as a total construction costs, but only construction works and components related to energy performance are to be included in the cost (energy performance related construction cost included in the calculations). Such construction works and components are:

- thermal insulation (with cost implications to other structures)
- windows
- air handling units
- heat supply solutions (boilers, heat pumps etc.)

In the example, in all calculated cases an under floor heating system was considered, that was not included in the energy performance related construction cost. The effect of maintenance, replacement and disposal costs is required to be taken into account in Cost optimal regulation. However, in the example, sensitivity analyses showed only minor differences between calculated cases, and these costs were not taken into account to keep calculations as simple and transparent as possible. Labour costs, material costs, overheads, the share of project management and design costs, and VAT are essential to include in the energy performance related construction cost.

Global cost, the term of EN 15459 used in the cost optimal regulation (= life cycle cost), and NPV calculation have follow EN 15459 [7]. Global energy performance related cost has to be calculated as a sum of the energy performance related construction cost and discounted energy costs for 30 years, including all electrical and heating energy use. Because the basic construction cost was not included, the absolute value of the global energy performance related cost will have a little meaning. Instead of that, the global incremental energy performance related cost was used. This can be calculated relative to the BAU construction:

$$C_g = \frac{C_I + \sum_{i=1}^{30} \left(C_{a,i} \cdot R_d(i) \right)}{A_{\text{floor}}} - \frac{C_g^{\text{ref}}}{A_{\text{floor}}} \tag{3}$$

where:

C_g Global incremental energy performance related cost included in the calculations, NPV, €/m^2

C_I Energy performance related construction cost included in the calculations, €

$C_{a,i}$ Annual energy cost during year i, €

$R_d(i)$ Discount factor for year i

C_g^{ref} Global energy performance related cost incl. in the calculations of BAU reference building, NPV, €

A_{floor} Heated net floor area, m^2

To calculate the discount factor R_d, real interest rate R_R depending on the market interest rate R and on the inflation rate R_i (all in per cents) is to be calculated:

$$R_R = \frac{R - R_i}{1 + R_i/100} \tag{4}$$

Discount factor for year i is calculated:

$$R_d(i) = \left(\frac{1}{1 + R_R/100} \right)^i \tag{5}$$

where:

R_R The real interest rate, %

I Number of a year, -

Global incremental cost calculation is illustrated in Table 3 for one case. A global incremental cost is negative if BAU is not cost optimal, and positive if the case studied leads to higher global cost than BAU.

To calculate the global energy performance related costs, the real interest rate and escalation of energy price has to be selected on national bases. In the example, the real interest rate of 3 % and escalation of energy prices of 2 % are used as basic case. Cost optimal regulation provides long term price development data for main fuels (oil, coal, gas) which can be utilized when estimating national energy price developments.

Table 3 Global incremental cost calculation

	DH 0.42	DH 0.58	DH 0.76	**DH 0.96 (ref.)**
Global energy performance related cost included in the calculations NPV, €				
Building envelope (thermal insulation and windows, structures not incl.)	30602	26245	21167	17611
Ventilation units (ductwork not included)	5474	3445	3445	3445
Condensing gas boiler (distribution system not included)	6917	6917	6917	6917
Solar collectors 6 m^2	4479	4479	0	0
Connection price: Gas	2455	2455	2455	2455
Energy cost for natural gas, NPV	10100	14063	22208	26196
Energy cost for electricity, NPV	20081	20081	20407	21422
Global cost included in the calculations, NPV, €	**80108**	**77685**	**76599**	**78047**
Global incremental energy performance related cost included in the calculations, relative to the reference building, NPV, €/m^2				
Building envelope (thermal insulation and windows, structures not incl.)	75.9	50.5	20.8	0.0
Ventilation units (ductwork not included)	11.9	0.0	0.0	0.0
Condensing gas boiler (distribution system not included)	0.0	0.0	0.0	0.0
Solar collectors 6 m^2	26.2	26.2	0.0	0.0
Connection price: Gas	0.0	0.0	0.0	0.0
Energy cost for natural gas, NPV	−94.1	−70.9	−23.3	0.0
Energy cost for electricity, NPV	−7.8	−7.8	−5.9	0.0
Global incremental cost included in the calculations, NPV, €/m^2	**12.0**	**−2.1**	**−8.5**	**0.0**

Global energy performance related cost included in the calculations is divided by net heated floor area of 171 m^2 and the values of the reference building (DH 0.96) are subtracted in order to calculate the global incremental cost. The global cost data shown corresponds to the "Gas" case in Fig. 2

Step 7. *Sensitivity analyses*

It is required in Cost optimal regulation to test at least the sensitivity to the real interest rate and energy prices. This will mean the calculation with lower and higher values.

3 Example: Estonian Reference Detached House

Global incremental energy performance related costs included in the calculations is shown in Fig. 2 for discounting interest rate of 1 % that corresponds to real interest rate of 3 % and escalation of 2 %. The global incremental cost is therefore presented as relative to the BAU construction concept DH 0.96 with gas boiler, that is very close to Estonian minimum requirement from 2008 regulation.

The results show two cost optimal values, as the construction concept DH 0.76 with gas boiler or ground source heat pump achieved the lowest NPV of the global incremental cost with marginal difference less than 2 €/m² NPV between these two heating systems. Negative NPV values compared to BAU show that the better

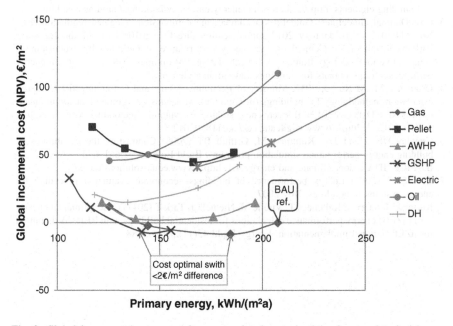

Fig. 2 Global incremental energy performance related costs in the reference detached house calculated with the real interest rate of 3 % and the escalation 2 %, and 30 years time period. (*AWHP* Air to water heat pump, *GSHP* Ground source heat pump, *DH* District heating.) For each heating system curve, the dots from *left* to *right* represent DH 0.42, 0.58, 0.76 and 0.96 construction concepts. The cost optimal values marked with arrows show that marginal, 2 €/m² change in the global cost led to highly significant change in the primary energy of about 40 units

construction standard can save some global cost. The lowest NPV defines the cost optimal performance level which is achieved for DH 0.76 construction concept with primary energy of about 180 kWh/(m² a) for gas boiler and about 140 kWh/(m² a) for ground source heat pump. As the global cost is less than 2 €/m² higher for ground source heat pump, the primary energy value of it would be relevant to select for the cost optimal energy performance level. This primary energy of 140 kWh/(m² a) is also achievable with reasonable global cost increase with air to water heat pump, gas boiler and district heating. More detailed description of Estonian cost optimal calculations can be found in [8].

References

1. EPBD: Directive 2010/31/EU of the European parliament and of the council of 19 May 2010 on the energy performance of buildings (recast). http://ec.europa.eu/energy/efficiency/buildings/buildings_en.htm, http://eur-lex.europa.eu/JOHtml.do?uri=OJ:L:2010:153:SOM:EN:HTML
2. Cost Optimal: Commission delegated regulation (EU) No. 244/2012 of 16 January 2012 supplementing directive 2010/31/EU of the European Parliament and of the Council on the energy performance of buildings by establishing a comparative methodology framework for calculating cost-optimal levels of minimum energy performance requirements for buildings and building elements. http://ec.europa.eu/energy/efficiency/buildings/buildings_en.htm
3. Cost Optimal Guidelines: Guidelines accompanying commission delegated regulation (EU) No: 244/2012 of 16 January 2012 supplementing directive 2010/31/EU of the European Parliament and of the Council on the energy performance of buildings by establishing a comparative methodology framework for calculating cost-optimal levels of minimum energy performance requirements for buildings and building elements
4. Ordinance MKM No 63: 2012, Ministry of Economic Affairs and Communications, Energy calculation methodology for buildings (in Estonian), Majandus- ja kommunikatsiooniministri määrus Nr. 63 (08 Oct 2012). Hoonete energiatõhususe arvutamise metoodika. Riigi Teataja, I, 18 Oct 2012, 1. https://www.riigiteataja.ee/akt/118102012001
5. Crawley DB, Hand JW, Kummert M, Griffith BT (2008) Contrasting the capabilities of building energy performance simulation programs. Build Environ 43:661–673
6. IDA-ICE, IDA indoor climate and energy 4.1. http://www.equa-solutions.co.uk/
7. EN 15459 (2007) Energy performance of buildings—economic evaluation procedure for energy systems in buildings, Nov 2007
8. Kurnitski J, Saari A, Kalamees T, Vuolle M, Niemelä J, Tark T (2011) Cost optimal and nearly zero (nZEB) energy performance calculations for residential buildings with REHVA definition for nZEB national implementation. Energy Build 43(11):3279–3288

Target Values for Indoor Environment in Energy-Efficient Design

Olli Seppänen and Jarek Kurnitski

Abstract This chapter summarises the factors which should be considered in the design and operation, focusing mainly on room temperature, indoor air quality, ventilation, moisture and humidity, noise and lighting. The information is mainly based on the European Standard EN 15251:2007, but as this standard does not cover all indoor environmental factors to be considered in low-energy building design also other sources are used. Many certification systems for buildings include in the evaluation criteria both energy use and the quality of the indoor environment. Thus, it is desirable to develop systems and solutions which lead to high-quality indoor environment with low energy use. The designer shall always document design criteria for the indoor environment; these criteria shall be available with the energy use data when renting or selling the building space. It is also recommended that design values for the indoor environment and indicators for the environmental comfort are included in the energy certificate and displayed with actual values for the energy use. This chapter describes also the difference between target values for dimensioning of systems and energy calculations. Different approaches for mechanically cooled buildings and buildings without mechanical cooling are introduced, and precautions are given for the latter if to be applied in low-energy buildings.

O. Seppänen (✉) · J. Kurnitski
Federation of European Heating, Ventilation and Air-conditioning Associations,
40 rue Washington, 1050, Brussels, Belgium
e-mail: oseppanen@rehva.eu

J. Kurnitski (ed.), *Cost Optimal and Nearly Zero-Energy Buildings (nZEB)*,
Green Energy and Technology, DOI: 10.1007/978-1-4471-5610-9_5,
© Springer-Verlag London 2013

1 Effects of Indoor Environment

Calculated energy consumption of buildings depends on the used criteria for the indoor environment by building and systems design and operation. Indoor environment affects also health, productivity and comfort of the occupants. Actually, the costs of the bad indoor environment for the society, employer and building owner are often higher than the cost of energy used in the same buildings. These effects are discussed in detail in the REHVA Guidebook No. 6 [10]. Energy use of buildings can be reduced dramatically if buildings are not heated, cooled or lit. This means that an energy declaration without relating it to the indoor environment makes no sense. Energy efficiency of buildings should not sacrifice comfort and health of occupants. There is therefore a need to specify criteria for the indoor environment for design, energy calculations, performance evaluation and display of operation conditions.

The information in this chapter is applicable in the non-industrial buildings where the criteria for indoor environment are set by human occupancy and where the production or process does not have a major impact on indoor environment. The values and recommendations are thus applicable to single-family houses, apartment buildings, offices, educational buildings, hospitals, hotels and restaurants, sports facilities, and wholesale and retail trade service buildings.

There exist several international standards and guidelines [1–4, 6], which specify criteria for indoor environment in different classes. Here, the recommended criteria are given for three classes (categories) I–III. Additional category IV is used for existing buildings, if indoor climate parameters do not meet category III requirement. Also for so-called free-running buildings without mechanical cooling, less strict criteria are used.

Using a higher class with stricter criteria may result in higher calculated design loads and then may result in larger systems and equipment. Applied values from a specific class may have also an influence on the energy demand. More stringent class typically means also higher occupant satisfaction and better comfort.

Selection of the category is building specific, and the needs of special occupant groups such as elderly people (low metabolic rate and impaired control of body temperature) shall be considered. For this group of people, category I is recommended. Category II is often used in new buildings as a default choice if no special design targets specified, because it is achievable with many technical solutions and allows enough variation in indoor climate parameters beneficial for energy-efficient design utilising both active and passive measures. It is also possible to select some parameters from category I (e.g. ventilation rate, air velocity) and indoor temperatures and others from category II, depending on design targets of the specific building. Category III cannot be recommended for new buildings due to potential harm to occupants, but can be used to classify the indoor environment in older buildings. If Category III requirements are not met, repairs are necessary to avoid the problems with indoor environment. Some descriptions on the effects of different categories are shown in Table 1.

Table 1 Effect of different categories of indoor environment on the human responses

Category of indoor environment	Room temperature	Ventilation rate	Draught
I—Excellent	$PPD^a < 6\ \%$	$PD^b < 15\ \%$	$DR^c < 10\ \%$ Only a few complaints by sensitive people on draught in the low indoor temperatures (winter)
II—Good	$PPD < 10\ \%$ Slightly reduced performance in work	$PD < 20\ \%$ Slightly increased rate of SBS-symptoms[d] Slightly reduced performance in work	$DR < 20\ \%$ Some complaints on draught in the low indoor temperatures (winter)
III—Satisfactory	$PPD < 16\ \%$ Significantly reduced performance in work Increased complaints on dry air and SBS-symptoms in winter Reduced productivity in text processing due to reduced dexterity of hands in winter	$PD < 30\ \%$ Increased rate of SBS-symptoms Reduced performance in work Risk of high humidity and microbial growth	$DR < 30\ \%$ Increased rate of complaints on draught in the low indoor temperatures (winter) Reduced productivity in text processing due to reduced dexterity of hands if hands are exposed to draught
IV—Poor	$PPD > 16\ \%$ Significantly reduced performance in work Significantly increased complaints on dry air and SBS-symptoms in winter Reduced productivity in text processing due to reduced dexterity of hands in winter	$PD < 30\ \%$ Significantly increased rate of SBS-symptoms Reduced performance in work Increased number of sick leaves in multiple occupant spaces Risk of high humidity and microbial growth	Significant increase rate of complaints on draught

(continued)

Table 1 (continued)

Category of indoor environment	Room temperature	Ventilation rate	Draught
Buildings without cooling	Significant dissatisfaction and reduced performance in low and high ends of the temperature range	Significant dissatisfaction and reduced performance in rooms with low ventilation rates	Potential high rate of complaints due to high air velocity in poorly designed buildings

[a] PPD predicted percentage of dissatisfied occupants with room temperature

[b] PD predicted percentage of dissatisfied visitors in room with perceived air quality

[c] DR draught rating, in this table, estimated percentage of people dissatisfied due to draught, in winter design conditions

[d] SBS sick building syndrome symptom = unspecific irritation of mucous membrane, skin, eyes, nose, fatigue, etc.

2 Indoor Temperature

2.1 Effects of Indoor Temperature

In many buildings, thermal conditions are not well-controlled due to insufficient cooling or heating capacity, high internal or external loads, large thermal zones, improper control system design or operation and other factors. Thermal conditions inside buildings may vary considerably with time, e.g., as outdoor conditions change, and spatially within buildings. While the effects of temperature on comfort are broadly recognised, indoor temperature could influence also productivity, learning and other activities.

Air temperature affects also air quality. Studies have shown that warm and humid air is stuffy and warm room air temperature in the winter causes a higher number of typical sick building symptoms than cooler air. These findings suggest the use of low room air temperature and low relative humidity in the winter from a standpoint of good indoor air quality (IAQ) and energy economy.

2.2 Design Indoor Temperature for Dimensioning

For design of building and dimensioning of HVAC systems, the thermal comfort criteria (minimum room temperature in winter, maximum room temperature in summer) shall be used as input for heating load and cooling load calculations. This will guarantee that a minimum–maximum room temperature can be obtained at design outdoor conditions and design internal loads. The recommended temperature criteria are given for three classes (categories). Using a higher class with stricter criteria will result in higher calculated design loads and then may result in larger systems and equipment. The designer shall document design criteria for the indoor environment. Some examples of design indoor operative temperatures criteria for different types of spaces are given in Table 1 for buildings with mechanical cooling. In most cases, the average room air temperature can be used as defining the design temperature, but if temperatures of large room surfaces differ significantly from the air temperature, the operative temperature should be used.

For buildings and spaces were the mechanical cooling capacity is not adequate to meet the required temperature categories, the design documents must state, how often the conditions are outside the required range.

The indoor temperatures are based on thermal comfort criteria for heated and mechanically cooled buildings [3]. Assuming different criteria for the PPD–PMV, each category of the indoor environment is established. For an assumed combination of activity and clothing, an assumed relative humidity and low air velocities, it is possible to establish a corresponding range of operative temperatures. For the design and dimensioning, further criteria for the thermal environment

Table 2 Examples of recommended design values of the indoor temperature for design of buildings and HVAC systems (from [2])

Type of building/space	Category	Operative temperature, °C	
		Minimum for heating (winter season), ~1.0 clo	Maximum for cooling (summer season), ~0.5 clo
Residential buildings: living spaces	I	21.0	25.5
(bed rooms, drawing room, kitchen, etc.)	II	**20.0**	**26.0**
Sedentary activity ~1.2 met	III	18.0	27.0
Single and open-plan offices, and spaces	I	21.0	25.5
with similar activity (conference rooms,	II	**20.0**	**26.0**
auditorium, cafeteria and restaurants)	III	19.0	27.0
Sedentary activity ~1.2 met			
Classroom	I	21.0	25.0
Sedentary activity ~1.2 met	II	**20.0**	**26.0**
	III	19.0	27.0

(draught, vertical air temperature differences, floor temperature and radiant temperature asymmetry) should be taken into account (see later).

The value of design temperature can vary from the values shown in Table 2 to take account of, e.g., local custom or a desire for energy saving so long as the within-day variation from the design temperature is within the given range, and the occupants are given time and opportunity to adapt to the modified design temperature.

2.3 Indoor Temperature for Energy Calculations

Standardised input values for the energy calculations are needed to perform a yearly energy calculation. Criteria for the indoor environment must be specified and documented.

For seasonal and monthly calculations, the same values of indoor temperature as for design (sizing) the heating and cooling systems should be used (Table 2) for each category of indoor environment to calculate energy consumption for heating and cooling, respectively.

In dynamic simulation, the energy consumption is calculated on an hourly basis. Recommended values for the acceptable range of the indoor temperature for heating and cooling are presented in Table 3. The mid-point of the temperature range should be used as a target value, but the indoor temperature may fluctuate within the range due to the energy saving features or control algorithm.

Table 3 Temperature ranges for calculation of cooling and heating energy in three categories of indoor environment (from [2])

Type of building or space	Category	Temperature range for heating, °C Clothing ~ 1.0 clo	Temperature range for cooling, °C Clothing ~ 0.5 clo
Residential buildings, living spaces	I	21.0–25.0	23.5–25.5
(bed rooms, living rooms, etc.).	**II**	**20.0–25.0**	**23.0–26.0**
Sedentary activity ~ 1.2 met	III	18.0–25.0	22.0–27.0
Single and open-plan offices and spaces	I	21.0–23.0	23.5–25.5
with similar activity (conference rooms,	**II**	**20.0–24.0**	**23.0–26.0**
auditorium, cafeteria, restaurants and classrooms). Sedentary activity ~ 1.2 met	III	19.0–25.0	22.0–27.0

2.4 Indoor Temperature in Buildings Without Mechanical Cooling

For the dimensioning of the heating system, the same criteria as for mechanically ventilated, cooled and heated buildings shall be used (Table 2).

The criteria for the thermal environment in buildings without mechanical cooling may be specified different from those with mechanical cooling during the warm season due to the different expectations of building occupants and adaptation. The level of adaptation and expectation in free-running buildings is strongly related to climatic conditions.

As there is no mechanical cooling system to dimension, the criteria for the categories of summer temperatures are mainly used for building design to prevent the overheating of the building by using solar shading, thermal capacity of building, design, orientation and opening of windows, etc. Based on a mean outdoor running mean temperature-recommended criteria for the indoor temperature are given in Fig. 1. The outdoor running mean temperature is the weighted average of mean average outdoor temperature of seven previous days (see [2] for exact weighting factors of outdoor running mean temperature).

The operative temperatures (room temperatures) presented in Fig. 1 are valid for office buildings and other buildings of similar type used mainly for human occupancy with mainly sedentary activities and dwelling, where there is easy access to operable windows and occupants may freely adapt their clothing to the indoor and/or outdoor thermal conditions. Dress code cannot be required in such buildings.

It has to be taken into account that the criteria of Fig. 1 are based on field studies in free-running buildings without mechanical cooling. When applied to new buildings, the expectations of occupants may be different compared to existing buildings with poor conditions. Therefore, the criteria may result in occupant dissatisfaction if applied for modern low-energy buildings. During the periods of elevated indoor temperatures, the productivity may also be deteriorated.

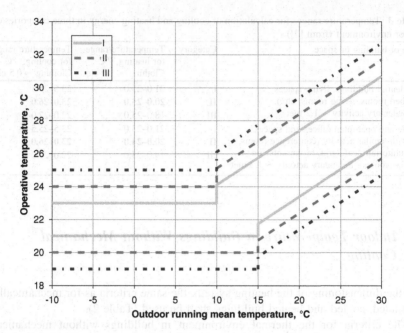

Fig. 1 Design values for the indoor operative temperature for buildings without mechanical cooling systems as a function of the exponentially weighted running mean of the outdoor temperature [2]. Temperature range for heating (*horizontal lines*) is shown for offices; the values are from Table 3

Another aspect making this criteria problematic in low-energy building design are large operable windows. As low-energy buildings need airtight building envelope and heat recovery in all European climates, such window solutions may not be feasible. For these reasons, the design with cooling according to criteria in Table 2 may result in improved energy performance and cost effectiveness, and better occupant satisfaction.

3 Local Thermal Discomfort

Criteria for local thermal discomfort such as draught, vertical air temperature differences and floor surface temperatures shall also be taken into account for the design of building and HVAC systems. A summary is given in Table 4 according to the Finnish criteria [6].

Table 4 Some criteria for local discomfort as specified in the Finnish classification of indoor environment

		Unit	Maximum values in three classes		
			S1	S2	S3
Air velocity	Winter (20 °C)	m/s	0.13	0.16	0.19
	Winter (21 °C)	m/s	0.14	0.17	0.20
Air velocity	Summer (24 °C)	m/s	0.20	0.25	0.30
Vertical temperature difference		°C	2	3	4
Floor temperature		°C	19–29	19–29	17–31

The classes S1–S3 used in Finland are similar to classes of [2], but provide some additional specification. The air velocity in the table is the omnidirectional average air velocity during 3 min in the occupied zone

3.1 Air Velocity and Draught

The air velocity in a space influences the convective heat exchange between a person and the environment. This influences the general thermal comfort of the body (heat loss) and the local thermal comfort due to draught (i.e. all air velocity effects cannot be explained by heat balance of the body as the skin is sensitive to air velocity that is stressed at lower temperatures and at higher turbulence intensity of air flow). There is no minimum air speed that is necessary for thermal comfort. Therefore, if within the comfort temperature range (Table 3), still air is desirable condition, however, not possible to reach because of convective (due to temperature differences) and air flows from supply air or room-conditioning devices. In any case, it is important to limit air velocity as much as possible (see Table 4), because draught complaints are one of the most common indoor climate complaints in office buildings, typically occurring during mid-season and winter when temperatures are reaching the lower limits of the comfort range.

In the cooling situation (indoor temperature reaching or above the upper limit of the summer range), increased air speed may be used to offset the warmth sensation caused by increased temperature during the warm season to increase the cooling effect of the body (note that this contradictory to the low velocities beneficial during the heating season to reduce body heat loss and special equipment is required to enable personal control of air speed in cooling situation).

Under "summer comfort conditions" (indoor operative temperatures >25 °C) increased air velocity may be used to compensate for increased air temperatures. Where there are fans (that can be controlled directly by occupants) or other means for personal air speed adjustment (e.g. personal ventilation systems), the upper limits presented in Fig. 2 can be increased by a few degrees. The exact temperature correction depends upon the air speed that is generated by the fan and can be derived from Fig. 2. This method can also be used to overcome excessive temperatures in mechanically controlled buildings if the local method for controlling air movement (fan, etc.) is available.

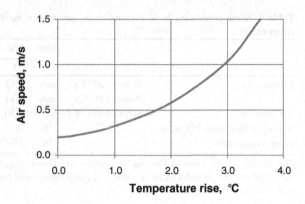

Fig. 2 Air speed required to offset increased temperature [3]. The air speed increases by the amount necessary to maintain the same total heat transfer from the skin. Acceptance of the increased air speed will require occupant control of device creating the local air speed [2]

3.2 Vertical Temperature Difference

The vertical temperature difference is the temperature difference between the ankle and neck level. The measuring heights are 0.1 and 1.1 m (sedentary work). It may cause discomfort even if the average room temperature in occupied zone is within acceptable range. The vertical temperature difference may become too large in the systems floor cooling, air heating of displacement ventilation.

3.3 Floor Temperature

Cold or hot feet can cause significant discomfort in floor heating or floor cooling systems, but also in the systems where thermal mass of the floor is used as heat storage either for heating or cooling. Floor temperature can be also too low or high if the poorly insulted floor is exposed to the outdoor conditions. The floor temperature anywhere in the occupied zone shall not be higher or lower than the temperature range in item "Floor temperature" in Table 4. In the bathroom, the maximum recommended floor temperature is 27 °C.

4 Air Quality and Ventilation

4.1 Health Effects of Indoor Air Pollutants

People spend 60–90 % of their life indoors—be it at home or in other public or private indoor environments, such as schools, cafés and restaurants. Having clean air indoors is very important for the health of the population as a whole, and it becomes particularly important for vulnerable groups like babies, children and the elderly or people already suffering from, e.g., respiratory or allergic diseases. The health effects of "cocktails" of different indoor pollutants, their concentrations and their public health significance are being studied worldwide. Already today, for many pollutants, scientific evidence shows a serious impact on the health of the population. Various indoor air pollutants are responsible for or exacerbate respiratory diseases, allergies, intoxication and certain types of cancer (e.g. asbestos, radon, environmental tobacco smoke (ETS), combustion products, volatile organic compounds, biological pollutants).

Because of the continuous air exchange, i.e., replacement of spent indoor air with fresh outdoor air, indoor air quality depends largely on outdoor air quality, but it depends also on a number of other variables including emissions form the building and its equipment, such as construction and surfacing materials, furnishings, heating and ventilating equipment, emissions from the use of consumer products for cleaning, preparation of food, and other occupant actions, e.g., smoking, opening/closing of windows as well as various hobbies and daily activities. Table 5 summarises the typical and high end levels of some indoor air contaminants and the contributions of the indoor sources to both the typical and the high end indoor air exposure levels in mainly west European conditions and compares the levels to the WHO (I)AQ Guidelines.

There is no common standard index for the indoor air quality but indoor air has to meet the WHO criteria for outdoor air pollutants. It is most likely, however, that WHO criteria for pollutants are not enough for good indoor air quality.

Limit values for indoor air pollutants can be used when checking in a building if the indoor air quality criteria are met. But they are very difficult to use in design work as the verified design tools for calculation of indoor air quality and ventilation are not available, and in addition, the input date of the emission of pollutants is not yet available for calculations. For this reason, the indoor quality and ventilation design are most commonly based on the selection of ventilation airflows.

Table 5 Typical and high-end levels of some indoor air contaminants and the contributions of the indoor sources to both the typical and the high-end indoor air exposure levels in Europe [9], and comparison to WHO (I)AQ guidelines [13, 14]

Agent	Long-term (I)AQG ($\mu g/m^3$)	Typical ($\mu g/m^3$)	Indoor source (%)	High end ($\mu g/m^3$)	Indoor source (%)
PM2.5	10	10–40	up to 30	100–300	>90
CO	10	1–4	0	100–200	>99
NO$_2$	40	10–50	up to 20	100–200	>75
Formaldehyde	30	20–80	>90	200–800	>99
Benzene	5	2–15	up to 40	50	>75
Naphthalene	10	1–3	up to 30	−1,000	>99.9
Radon (Bq/m^3)	200	20–100	>90	−100,000	>99.9

4.2 Health Effects of Ventilation

Ventilation in buildings is intended to remove pollutants and reduce their concentration, and control thermal conditions. Ventilation has complex effects on
indoor air quality, health, performance, comfort, dampness and airflows in
buildings. Exposure to pollutants in indoor air and ventilation may cause a variety
of effects. The severity of the effects covers a wide spectrum from perception of
malodours to cancer. The effects may be acute or develop over a longer period of
time. Literature has shown that ventilation is associated with the health, but the
exact dose–response relation cannot yet be established due to the large variability
in pollutants and conditions. As the limit values and source strengths are not
known for all pollutants, the exact determination of required ventilation rates
based on pollutant concentrations and associated risks is never possible. The
selection of ventilation rates has also to be based on epidemiological research,
laboratory and field experiments, odour perception, irritation, occupant preference,
productivity and experience.

The published papers indicate that ventilation rates can be kept as low as 7 l/s
per person while no elevated risk of asthma and allergic symptoms are identified,
and as low as 8–9 l/s person so that no increase in the onset of subjectively
reported symptoms related to the presence in the building can be registered,
summarised by latest review in European HealthVent project [7, 8].

The evidence suggests that better hygiene, commissioning, operation and
maintenance of air-handling systems may be important for reducing the negative
effects of HVAC systems.

4.3 Ventilation Rates and Emission of Pollutants

In the design and operation, the main sources of pollutants should be identified and
eliminated or decreased by any feasible means. Local exhausts and ventilation then
deal with the remaining pollution. Air-cleaning devices can also be used to remove
the pollutants from the room air to improve the air quality. Ventilation should be
used when the source control is not possible (e.g. human occupancy) or is not
feasible or is too expensive. In theory, source control is easier to implement and
more effective at the different levels of the sources.

4.4 Indoor Air Quality and Ventilation Rates

For design of ventilation systems and calculation of heating and cooling loads, the
required ventilation rate must be specified in the design documents based on
national requirements or using the recommended methods in the standards. It is

generally accepted that the indoor air quality is influenced by emissions from people and their activities and from building and furnishing, and from the HVAC system itself. The two last sources are normally called the building components. In the standard EN 15251 [2], the recommended ventilation rates in non-residential buildings are derived taking into account pollutant emission. The calculated design ventilation rate is from two components (a) ventilation for pollution from the occupancy and (b) ventilation for the pollution from the building itself. The ventilation for each category is the sum of these two components as illustrated with Eq. 1, which is to be used to calculate the total ventilation rate for a room:

$$q_{\text{tot}} = n \cdot q_p + A \cdot q_B \qquad (1)$$

where

q_{tot} total ventilation rate of the room, l/s

n design value for the number of the persons in the room

q_p ventilation rate for occupancy per person, l/s, pers

A room floor area, m^2

q_B ventilation rate for emissions from building, l/s, m^2

The ventilation rates for occupants (q_p) only and the ventilation rates (q_B) for the building emissions are shown in Table 6.

Low-polluting and very-low-polluting building materials have to meet strict criteria specified in EN 15251 [2] for material emissions and odours. In some countries, material labelling systems and labelled materials are available corresponding to very-low-polluting material, the Finnish M1 being the best-known example and available for most wide product range.

Examples of the total ventilation rates for non-industrial, non-residential buildings based on these values are calculated using Eq. 1 with default occupancy densities (Floor area m^2/person) indicated in Table 7.

Criteria for the ventilation rate may also be expressed as total rates per m^2 floor area (l/s, m^2) or per occupant l/s per occupant. By expressing it as a people part and as a building part, it will be easier to calculate required ventilation rates for non-typical level of occupancies.

In most cases, the health criteria will also be met by the required ventilation for comfort (i.e. ventilation rates in Table 7). Health effects may be attributed to specific components of emission, and if you reduce concentration of one source, you also reduce concentration of others. Comfort is more related to the perceived

Table 6 Ventilation rate components to be used to calculate the total ventilation rate of the room with Eq. 1 [2]

	q_p, occupants only, l/s, pers	q_B, very-low-polluting building, l/s, m^2	q_B, low-polluting building, l/s, m^2	q_B, non-low-polluting building, l/s, m^2
Category I	10	0.5	1.0	2.0
Category II	7	0.35	0.7	1.4
Category III	4	0.3	0.4	0.8

Table 7 Recommended ventilation rates for non-residential buildings with default occupant density for two categories of pollution from building itself [2]

Type of building or space	Category	Floor area $m^2/$ person	q_p l/s, m^2 for occupancy	q_B l/s, m^2 for low-polluted building	q_B l/s, m^2 for non-low-polluted building	q_{tot} l/s, m^2 total for low-polluted building	q_{tot} l/s, m^2 total for non-low-polluted building
Single office	I	10	1.0	1.0	2.0	2.0	3.0
	II	10	0.7	0.7	1.4	1.4	2.1
	III	10	0.4	0.4	0.8	0.8	1.2
Landscaped office	I	15	0.7	1.0	2.0	1.7	2.7
	II	15	0.5	0.7	1.4	1.2	1.9
	III	15	0.3	0.4	0.8	0.7	1.1
Conference room	I	2	5.0	1.0	2.0	6.0	7.0
	II	2	3.5	0.7	1.4	4.2	4.9
	III	2	2.0	0.4	0.8	2.4	2.8
Classroom	I	2.0	5.0	1.0	2.0	6.0	7.0
	II	2.0	3.5	0.7	1.4	4.2	4.9
	III	2.0	2.0	0.4	0.8	2.4	2.8
Kindergarten	I	2.0	6.0	1.0	2.0	7.0	8.0
	II	2.0	4.2	0.7	1.4	4.9	5.8
	III	2.0	2.4	0.4	0.8	2.8	3.2

air quality (odour, irritation). In this case, different sources of emission may have an odour component that adds to the odour level. There is, however, no general agreement how different sources of emission should be added together.

Attention has to be paid when ventilation rates for very-low-polluted buildings are intended to use, because depending on occupant density, these may stay below the limit values associated with increased sick leaves or negative effects on performance. According to HealthVent summary, at least 15 l/s per person in offices and 7 l/s per person in schools are required to avoid these negative effects. Therefore, it is good to check that the total ventilation rate q_{tot} calculated with Eq. 1 (and to be multiplied with floor area m^2/person in order to get ventilation rate l/s per person) will be at least equal to these limit values.

4.5 Ventilation in Residential Buildings

Indoor air quality in residential buildings depends on many parameters and sources such as number of persons (time of occupation), emissions from activities (smoking, humidity, intensive cooking), emissions from furnishing, flooring materials and cleaning products, and hobbies. Humidity is of particular concern in residential ventilation as most of the adverse health effects and building disorder (condensation, moulds,) are related to humidity. Several of these sources cannot be influenced or controlled by the designer.

Required design ventilation rates shall be specified as an air change per hour for each room, and/or outside air supply and/or required exhaust rates (bathroom, toilets and kitchens) or given as an overall required air change rate. Most national regulations and codes give precise indications on detailed airflows per room and shall be followed. The required rates shall be used for designing mechanical, natural and exhaust ventilation systems.

Residential ventilation can be based at least on the following three criteria:

- Exhaust of pollutions in "wet" rooms (bathroom, kitchen, toilets).
- General ventilation of all rooms in the dwelling (the total volume).
- General ventilation of all rooms in the dwelling with ventilation criteria in the main rooms (bed and living rooms).

The default ventilation rates in Table 8 are based on average use of a residence. In operation, some residences may need more ventilation and some may manage with lower ventilation rates. National regulations as well as international standards help the designer to determine assumptions made on standard residential sources and the correct airflow to achieve.

Example of procedure for selecting the ventilation rate:

When the values of any specific category in the table lead different values of the ventilation depending on the number of occupants, floor area and number of kitchen, bathroom and toilet exhausts, the following principle should be followed:

1. Calculate total ventilation rate for the residence based on

   (a) Floor area, column (1) and
   (b) Number of occupants or number of bedrooms, column (2) [if the number of occupants is not known use column (3)].

Table 8 Example of ventilation rates for the residences

Category	Air change rate of the whole residence[a]		Living room and bedrooms, outdoor air flows		Exhaust air flow, l/s		
	l/s, m² (1)	ach	l/s, pers[b] (2)	l/s/m² (3)	Kitchen (4a)	Bathrooms (4b)	Toilets (4)
I	0.49	0.7	10	1.4	28	20	14
II	0.42	0.6	7	1.0	20	15	10
III	0.35	0.5	4	0.6	14	10	7

Continuous operation of ventilation during occupied hours. Complete mixing
[a] The air change rates expressed in l/s m² and ach correspond to each other when the ceiling height is 2.5 m
[b] The number of occupants in a residence can be estimated from the number of bedrooms. The assumptions made at national level have to be used when existing, and they may vary for energy and for IAQ calculations

2. Select the higher value from above (a) or (b) for the total ventilation rate of the residence.
3. Adjust the exhaust air flows from the kitchen, bathroom and toilets, columns (4) accordingly

   (a) in residences with small floor area exhaust air flow rates become smaller and
   (b) in large residences higher.

4. Outdoor air should be supplied primarily to living rooms and bedrooms.

The values in the table assume complete mixing in the room (i.e. concentration of pollutants is equal in exhaust and in occupied zone).

4.6 Evaluation of Ventilation Based on CO_2-Concentration

If the occupants are the only source of pollution in a building, the ventilation can be designed based on the CO_2 level. Table 9 gives the recommended values of CO_2 concentration to be used in the design.

4.7 Filtration and Air Cleaning

Although filtration is usually dimensioned for maintaining equipment performance, it can also be used to improve indoor air quality with:

- limiting the entry of particulate matter (particles from combustion, traffic, pollen, etc.) from outdoors.
- treatment of outdoor air in very polluted area.
- removal of odours and gaseous contaminants (gas-phase air cleaning).

Fine particle filters (F7 or F8 class) can be recommended for most of the locations with typical outdoor pollution as they provide effective protection of equipment as well as against fine particulate matter PM2.5 associated with mortality and heart illness. Design guidelines on air cleaning and filtration are given in EN 13779 [4] and in the REHVA Guidebook No. 11 [11].

Table 9 Recommended CO_2 concentrations above outdoor concentration for energy calculations and demand control [2]. Default outdoor concentration is 400 ppm (parts per million)	Category	Corresponding CO_2 above outdoors in ppm for energy calculations
	I	350
	II	500
	III	800
	IV	>800

5 Moisture and Air Humidity

5.1 Moisture in Buildings

Low ventilation may lead to high indoor humidity and moisture accumulation into building structures or materials. That may lead to increased dust mites, and particularly high humidity can increase the risk of microbial growth, and subsequently to microbial contamination and other emissions in buildings. In epidemiological studies, moisture damage and microbial growth in buildings have been associated with a number of health effects including respiratory symptoms and allergic diseases and other symptoms although the evidence of a direct link between higher air humidity levels and adverse health is quite limited. The health effects associated with moisture damage and microbial growth seem to be consistent in different climates and geographical regions.

The underlying mechanisms are not well understood because the specific agents that cause the health effects have not been identified with certainty. However, particles that come from mould and bacterial contaminants are likely to be the cause of these health effects. Also toxic mechanisms may possibly be involved, especially in connection with toxin-producing fungi and bacteria. The primary controlling factor of the mould growth is the level of moisture content within the building materials.

5.2 Air Humidity

Humidity has only a small effect on thermal sensation and perceived air quality in the rooms of sedentary occupancy; however, long-term high-humidity indoors (>60 %) may cause microbial growth, and very low humidity (<15–20 %) causes dryness and irritation of eyes and air ways. Requirements for humidity influence the design of dehumidifying (cooling load) and humidifying systems and will influence energy consumption. The criteria depend partly on the requirements for thermal comfort and indoor air quality and partly on the physical requirements of the building (condensation, mould, etc.). For special buildings (museums, historical buildings, churches), additional humidity requirements must be taken into account. Humidification or dehumidification of room air is usually not required in all European climates, but if used, excess humidification and dehumidification should be avoided. Some dehumidification typically happens in air-conditioning (cooling), and this can be limited with high-temperature cooling or desiccant cooling applications. Scientific evidence does not support humidification in winter time in buildings designed for human occupancy, and this is supported by at least 30 years of operation experience especially from Northern Europe, where humidification was stopped to use because of hygiene and energy considerations and any increase in adverse health effects has not been reported. If humidity

Table 10 Humidification or dehumidification is usually not needed, but if humidification or dehumidification systems are installed, example of recommended design criteria for the humidity in occupied spaces may be used [2]

Type of building/space	Category	Design relative humidity for dehumidification, %	Design relative humidity for humidification, %
Spaces where humidity criteria are	I	50	30
set by human occupancy. Special	II	60	25
spaces (museums, churches, etc.)	III	70	20
may require other limits	IV	>70	<20

control is decided to use, recommended design values of indoor humidity for occupied spaces for dimensioning of dehumidification and humidification systems are given in Table 10.

6 Lighting

6.1 Effect of Lighting on Health and Productivity

Natural daylight has a significant and positive influence on occupant health, well-being and productivity. However, adaptive control of daylight is needed to guarantee the conditions of good visual comfort at all times. Several examples are referred on positive effect of daylighting [12].

By maximising the use of daylight without glare and providing daylight-responsive lighting controls, a productivity benefit of between 0.45 and 40 % was found by Carnegie Mellon University. On average, major health complaints are between 20 and 25 % lower for persons close to an exterior window, compared to those that work in the interior core without access to view and daylight. Office workers were found to perform 10–25 % better on tests of mental function and memory recall when they had the best possible view versus those with no view. Direct sun penetration into classrooms, especially through unshaded east- or south-facing windows, was associated with negative student performance, likely causing both glare and thermal discomfort. Students with adequate natural daylight in their classrooms showed 20 % faster progress in math tests and 26 % in reading tests during one year.

6.2 Visual Comfort

There is no doubt that people prefer daylight to electric lighting as their primary source of light. Visual contact with the outside world is also generally recognised as an important factor influencing people's positive emotional states. Despite these positive aspects of windows and daylight, situations that cause visual discomfort can easily arise in a day-lit office. Occasionally, light is just too bright or contrasts are too large. To fully harvest the benefits of daylight, it needs to be regulated.

Luminance[1] is the physical quantity which most closely corresponds to what people call "brightness". Research shows that good visual comfort is experienced when the luminance within the central field of view is no more than three times the luminance of the visual task, and no less than one-third of it. Luminance within the peripheral field of view should be within 0.1 and 10 times the luminance level of the visual task. The idea is illustrated in Fig. 3.

Discomfort glare is caused by high luminance ratios within the field of view. Severe glare may disrupt work and may even cause physiological disorders. Glare is usually caused by direct sunlight falling on objects in the office or high luminance values in the exterior within the field of view. Glare can also occur when using a computer display. The luminance of the reflection of the surroundings may be higher than the luminance of the computer screen. Without sun shading to attenuate and diffuse direct sunlight, the conditions for good visual comfort are often violated.

Visual comfort is also affected by colour rendition. Colour rendition is determined by the spectral composition of the illuminating light source. Unfiltered natural daylight gives by far the best colour rendition.

Contact with the outdoors is an important aspect of visual comfort. Obviously, when lowered, solar shading will at least partially obstruct a view to the outdoors. The degree of obstruction is determined by the openness of the shading. Slatted devices may offer a view through depending on the slat angle. Smaller slat widths are generally preferred. Screen fabrics will generally have an openness factor of several percentages. This usually gives a reasonable view of the outdoors. Fabrics with a dark interior and a low-light transmittance through the fibres are to be preferred from this perspective. In that case, the luminance of the screen itself will be relatively low in comparison with the luminance of the exterior scene visible through the openings in the fabric.

6.3 Lighting in Non-residential Buildings

To enable people to perform visual tasks efficiently and accurately, adequate light (without side effects such as glare and blinding) must be provided. The design luminance levels can be secured by means of daylight, artificial light or a combination of both. For reasons of health, comfort and energy in most cases, the use of daylight (maybe with some additional lighting) is preferred over the use of

---

[1] Luminance is measured in cd/m^2 and is a property of extended (direct and indirect) light sources. Luminance is defined as the luminous power per unit area per unit solid angle. This is the luminous flux in lumen emitted by a small patch in a certain direction within a certain solid angle.

Fig. 3 Luminance ratios for good visual comfort. In case of artificial lighting (*above*) and in case of broad daylight (*below*) [12]

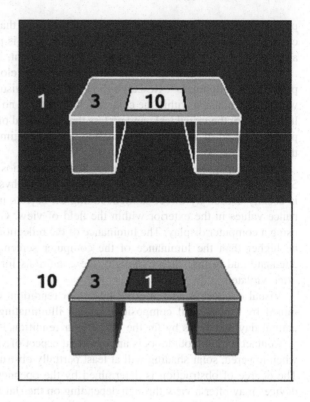

artificial light. The required task illuminance is defined and detailed in EN 12464-1 [5] and for some tasks is presented in Table 11.

The required lighting level is independent of season and the same criteria as for dimensioning of lighting systems shall be used for energy calculations. The required lighting level can be obtained by natural lighting, artificial lighting or a combination. The choice of light source will have a significant impact on the building energy demand. Energy for lighting is calculated only for the occupied hours based on the agreed occupancy profile. It is essential to evaluate also the quality of lighting in the energy calculations in respect of glare which may affect the use of controls and window screens. Recommended criteria for lighting are described in detail in EN 12464-1. Some of the criteria from standard are presented in Table 11.

7 Noise

The noise from the HVAC systems of the building may disturb the occupants and prevent the intended use of the space or building. The noise in a space can be evaluated using A-weighted equivalent sound pressure level. Table 12 gives

Table 11 Examples of design illumination levels for some buildings and spaces from EN 12464-1 [5]

Type of building	Space	Maintained luminance, $\hat{E}_m$, at working areas, lx	UGR[a]	Ra	Remarks
Office buildings	Single offices	500	19	80	At 0.8 m
	Open-plan offices	500	19	80	At 0.8 m
	Conference rooms	500	19	80	At 0.8 m
Educational buildings	Classrooms	300	19	80	At 0.8 m
	Classrooms for adult education	500	19	80	At 0.8 m
	Lecture hall	500	19	80	At 0.8 m
Circulation areas	Corridor	100	28	40	At 0.1 m
	Stairs	150	25	40	At 0.1 m
Other buildings	See EN 12464-1 [5]				

For information purposes, the UGR and Ra are also presented [2]

[a] *UGR* (unified glare rating) is used to evaluate the glare from artificial lighting

Table 12 Indoor system noise criteria of some spaces and buildings—examples A-weighted sound pressure level [2]

Building	Type of space	Sound pressure level [dB(A)]	
		Typical range	Default design value
Residential	Living room	25–40	33
	Bed room	20–35	28[a]
Offices	Small offices	30–40	35
	Conference rooms	30–40	35
	Landscaped offices	35–45	40
	Office cubicles	35–45	40
Schools	Classrooms	30–40	35
	Corridors	35–50	40
	Gymnasiums	35–45	40
	Teacher rooms	30–40	35
	Swimming baths	40–50	45

[a] Practise has shown that this is too high for HVAC equipment in residences and may cause the shutting down of the equipment, the Finnish guideline value is 24 dB(A) (Classification of Indoor Environment 2008)

typical values for some spaces as well as default design values. These criteria apply to the sources from the building as well as the noise level from outdoor sources. The criteria should be used to limit the sound power level from the mechanical equipment and to set sound insulation requirements for the noise from outdoors and adjacent rooms. Often national requirements exist for noise from outside, assuming that windows are closed.

The values can be exceeded in the case when the occupant can control the operation of the equipment or the windows. For example, a room air-conditioner may generate a higher sound pressure level if its operation is controlled by the occupant, but even in this case, the rise of the sound pressure level over the values in Table 12 should be limited between 5 and 10 dB(A).

Ventilation should not rely primarily on operable windows if the building is located in an area with a high outdoor noise level compared to the level the designer wishes to achieve in the indoor zone.

References

1. CEN Technical Report CR 1752 (2001) Ventilation for buildings—design criteria for the indoor environment
2. EN 15251 (2007) *European Standard*. Indoor environmental input parameters for design and assessment of energy performance of buildings—addressing indoor air quality, thermal environment, lighting and acoustics
3. EN ISO 7730. Analytical determination and interpretation of thermal comfort using calculation of the PMV and PPD indices and local thermal comfort
4. EN 13779 (2007) *European Standard*. Ventilation for non-residential buildings— performance requirements for ventilation and room-conditioning systems
5. EN 12464-1 Light and lighting—lighting of work places-part 1: indoor work places
6. FiSIAQ (2008) Classification of indoor environment 2008 target values, design guidance, and product requirements. Finnish Society of Indoor Air Quality and Climate
7. Health-based ventilation guidelines for Europe. Executive summary of HealthVent project. Project granted by the EU's Executive Agency for Health and Consumers, 2013
8. HealtVent Project Report WP4—Health and ventilation: review of the scientific literature, 20 Sept 2012. www.healthvent.eu
9. Jantunen M, Oliveira Fernandes E, Carrer P (2011) Universita degli studi di Milano, Stylianos Kephalopoulos, Promoting actions for healthy indoor air (IAIAQ), Final Report to DG Sanco. http://ec.europa.eu/health/healthy_environments/docs/env_iaiaq.pdf
10. REHVA Guidebook no. 6. Indoor climate and productivity in offices. Federation of European Heating, Ventilation and Air-conditioning Associations. www.rehva.eu
11. REHVA Guidebook no. 11. Air filtration in HVAC systems. Federation of European Heating, Ventilation and Air-conditioning Associations. www.rehva.eu
12. REHVA Guidebook no. 12. Solar shading. Federation of European Heating, Ventilation and Air-conditioning Associations. www.rehva.eu
13. WHO. Air quality guidelines for Europe. WHO Regional Publications, European Series, No. 91, 2000. Regional Office for Europe, Copenhagen
14. WHO. Air quality guidelines—Global update 2005. World Health Organisation, Regional Office for Europe, Copenhagen

Energy Efficiency Measures: In Different Climates and in Architectural Competitions

Panu Mustakallio and Jarek Kurnitski

Abstract Energy use of buildings is strongly affected by the climate the building is located. Some measures are effective in all climates, but attention to energy balance components and proper solutions depends on climate. An office building case study is used to show the performance in all climates, temperate, Mediterranean, cold and tropical described with Paris, Rome, Stockholm and Bombay weather data. It is shown that energy performance can be strongly improved with energy-efficient building envelope elements especially for windows and solar shading, modern lighting system with intelligent controls and optimal HVAC system with very efficient heat recovery, good chiller design and a high-temperature room-conditioning application. When building is located in Mediterranean or tropical climate conditions, significant part of energy use comes from cooling/drying of supply air, stressing the importance of corresponding solutions. Energy efficiency measures are evidently important design issues, to be tackled already in very early stages with integrated design. This applies also for architectural competitions. The problem is that if energy performance targets will be applied after architectural competition, this might be too late, and in worst case, the whole proposal has to be redesigned to meet the targets. To avoid such problems, energy performance targets are to be included in the competition brief among all other targets. It is discussed how energy performance targets can be included so that they will lead to integrated design from very first steps, but unnecessarily, complicated and detailed analyses can be avoided. Two possible approaches, one based on simple indirect indicators requiring a minimum calculation effort and another based on energy simulations, are discussed. A case study example with the application of the second approach is reported.

P. Mustakallio (✉) · J. Kurnitski
Halton Group, Helsinki, Finland
e-mail: panu.mustakallio@halton.com

J. Kurnitski (ed.), *Cost Optimal and Nearly Zero-Energy Buildings (nZEB)*,
Green Energy and Technology, DOI: 10.1007/978-1-4471-5610-9_6,
© Springer-Verlag London 2013

1 Energy Efficiency Measures in Different Climates

1.1 Technical Solutions and the Office Building Studied

To evaluate the energy efficiency and the energy-saving potential, three different kinds of buildings were used with same floor plan and window sizes (see Fig. 1). Basic reference building has structures, which has been typically used in Central European climate conditions and according to the local regulations for new buildings [1]. In advanced building, thermal conductivity of external structures has been improved as well as solar shading of window and lighting system. In low-energy building, the structures were the same as in the advanced building case, but lighting system was still improved. Energy simulations were done with IDA-ICE 4.0 tool by calculating the annual energy need of the office building. The simulation tool is validated according the International Energy Agency's validation exercises [2].

1.2 Basic Reference Building

Selection of indoor temperatures and ventilation rates were based on EN 15251 [5]. Energy-efficient heating, ventilation and air-conditioning (HVAC) systems were simulated for room air-conditioning and ventilation [3, 4]. The basic reference building was simulated with dedicated outdoor air system (DOAS), active chilled beams (with constant air volume) and fan coils where supply air and water were used for cooling/heating, and with variable air volume (VAV) system where only

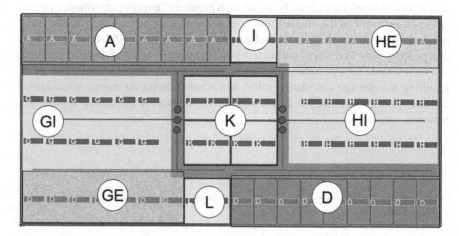

Fig. 1 Floor plan of the building and division to several operational zones: office rooms (*A, D*), landscape offices (*H, G*) and meeting rooms (*I, K, L*)

supply air was used for cooling/heating. Active chilled beams were selected for HVAC system in the basic reference building because of their lowest energy and electricity use. With fan-coil system, the fan efficiency of individual fans in the room units causes the bigger energy use, and with VAV system, the use of air as the media for cooling/heating power makes it less energy efficient in common office spaces where required supply air volume is not big. Cooling water for the air handling unit (AHU) and chilled beams was cooled by using one chiller system where the coefficient of performance (COP) was calculated with model taking into account part load ratios and outside air temperature of real chiller.

1.3 Advanced Building

For advanced building, chilled beam system was changed to more energy-efficient adaptable active chilled beams with VAV function for meeting room based on CO_2 concentration of room air. Also other HVAC system features were changed, like efficiency of heat recovery system and chiller design. Two chillers were used in advanced and low-energy buildings in order to get higher COP from the high-temperature cooling of chilled beam system and handle the AHU low-temperature cooling with other chiller. There was also added free cooling circuit so that the capacity of outside air temperature is used for cooling when possible. The chilled ceiling system (water circulated) was also simulated in the advanced building case in order to find out whether better energy efficiency could be achieved. The HVAC system selection for advanced building was still adaptable chilled beams, because the energy use was nearly the same with these systems, and for calculating the cases in different climate conditions, chilled beam system can provide more cooling power more flexibly.

1.4 Low-Energy Office Building

For low-energy office building case, the same chilled beam system was used as in advanced building case, but there was lower pressure level in the ventilation system. Also heat recovery was changed to a yet more efficient rotating wheel and lighting system to LED-based lighting with occupancy control.

1.5 Common Features of the Office Building Case

The simulation was made using 11,000-m^2 office building (10 floors), each floor with a mixture of different types of spaces: landscape offices 610 m^2 (55 %), office rooms 242 m^2 (22 %), meeting rooms 162 m^2 (15 %) and other (rest rooms, etc.)

Table 1 The heat load levels and schedules for occupancy, equipment, lighting and ventilation

	Basic reference building	Advanced building	Low-energy building
Occupants	Mo–Fri 8–18	Mo–Fri 8–18	Mo–Fri 8–18
Maximum number of occupants, m^2/person	10 (office), 15 (landscape), 2 (meeting room)	10 (office), 15 (landscape), 2 (meeting room)	10 (office), 15 (landscape), 2 (meeting room)
Average occupancy in offices, %	57.5	57.5	57.5
Average occupancy in meeting rooms, %	28.6	28.6	28.6
Equipment	Mo–Fri 8–18	Mo–Fri 8–18	Mo–Fri 8–18
Maximum equipment load, W/m^2	20 (office), 15 (landscape), 30 (meeting room)	20 (office), 15 (landscape), 30 (meeting room)	20 (office), 15 (landscape), 30 (meeting room)
Average equipment load ratio, %	Same as occupancy	Same as occupancy	Same as occupancy
Lighting	Mo–Fri 7–20	Mo–Fri 7–20	Mo–Fri 7–18
Lighting load, W/m^2	15	12	6
Control principle	Time	Time + daylight	Daylight + occupancy
Ventilation	Mo–Fri 7–19	Mo–Fri 7–19	Mo–Fri 7–19

95 m^2 (8 %). The main facades were towards north-west and south-east. Window height was 1.8 m and width 1.2 m, one window in each 1.35 m module, so window–floor ratio was 25 % in external offices. The heat load levels and schedules for occupancy, equipment, lighting and ventilation are presented in Table 1. They were typical for usual office building. Other building and system design parameters are presented in Table 2. Energy simulation was made using Paris–Orly weather data in all basic reference, advanced and low-energy cases. Then, the low-energy case was also simulated then with Stockholm, Rome and Bombay weather data to analyse the situation more comprehensively.

1.6 Energy Use

The total energy use and division for cooling, heating, ventilation fan energy, pumping and lighting are shown in the Fig. 2 as energy delivered to the building and in Fig. 3 as primary energy where gas for heating and electricity is weighted according to the efficiency of the energy production. Usual values for the primary energy factors have been used: 1 for gas and 2.5 for electricity.

The annual delivered energy use of the low-energy building, the most energy-efficient office building, in middle European climate is 22 kWh/m^2 and the primary energy use is 49 kWh/m^2. In the advanced building, the building with most common energy-efficient features, both delivered energy and primary energy uses are two times higher, and in the basic reference building with good standard construction, both uses are almost four times.

Table 2 Design values of reference, advanced and low-energy building

	Basic reference building	Advanced building	Low-energy building
External wall W/K, m²	0.43	0.3	0.3
Infiltration L/s, m²	0.33	0.165	0.165
Window W/K, m²	2.6	1.1	1.1
Window g-value	0.48	0.31	
Solar shading	No	External overhang of 500 mm	External overhang of 500 mm
Room terminal unit	Traditional active chilled beam	Adaptable active chilled beam	Adaptable active chilled beam
Chiller plant with air cooled condensers	One common chiller (2 °C)	2 chillers: beams (10 °C), AHU (2 °C)	2 chillers: beams (10 °C), AHU (2 °C)
Room unit inlet water temperature	15 °C	15 °C	15 °C
Room control	P	PI	PI
Room temperature set value	20.5/24 °C	20.5/25 °C	20.5/25 °C
Boiler plant	Condensing boiler	Condensing boiler	Condensing boiler
Airflow rates	DOAS: 1.5 L/s, m² in offices, 4.2 L/s, m² in meeting rooms	DOAS: 1.5 L/s, m² in offices, 4.2 L/s, m² in meeting rooms	DOAS: 1.5 L/s, m² in offices, 4.2 L/s, m² in meeting rooms
Airflow design	CAV in all spaces	CAV in offices, VAV in meeting rooms	CAV in offices, VAV in meeting rooms
Ductwork	Balanced	Constant pressure	Constant pressure
SFP kW/m³, s	2.0	1.8	1.2
AHU cooling water temperature	7 °C	7 °C	7 °C
Supply air temperature	16 °C (External $T > 20$ °C) 20 °C (External $T < 10$ °C)	16 °C (External $T > 20$ °C) 20 °C (External $T < 10$ °C)	16 °C (External $T > 20$ °C) 20 °C (External $T < 10$ °C)
Heat recovery	Hydronic (40 %)	Plate (60 %)	Wheel (80 %)
Filtration	EU7 in supply and EU3 in exhaust	EU7 in supply and EU3 in exhaust	EU7 in supply and EU3 in exhaust
Night purge ventilation	No	Yes	Yes

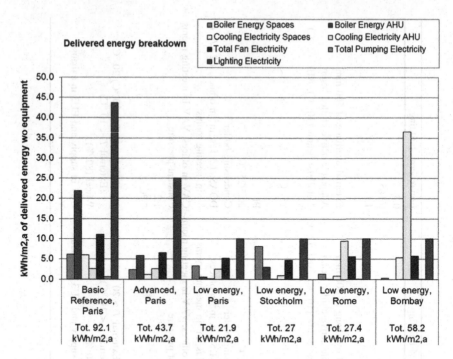

Fig. 2 Delivered energy use in different cases

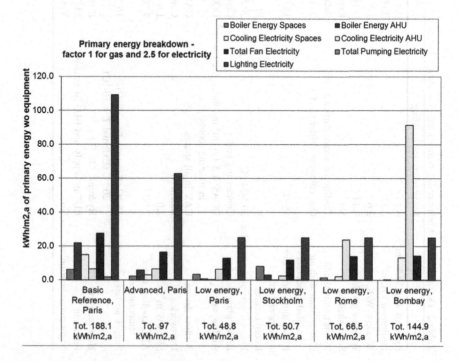

Fig. 3 Primary energy use in different cases

The biggest primary energy consumer is lighting, in the standard building four times bigger than second consumer and in the most efficient building in Paris two times. The second biggest is fan energy in these buildings. Heating and cooling energy demand is very small in the most efficient building. The cooling and heating energy breakdown is shown in the Figs. 4 and 5. The biggest reason for that is the efficient solar shading and very efficient heat recovery in AHU, which reduces significant amount of heating energy especially in the case of Nordic climate, and cooling energy in the case of Mediterranean and tropical climate.

Energy-efficient lighting system in the advanced and especially in low-energy buildings also lowers the internal heat load level so that the effect to the cooling energy use remains small. The primary energy for cooling and drying of ventilation supply air is the second biggest consumer in Mediterranean and clearly biggest in tropical climate.

The comparison of energy use in the office building with standard, advanced and low-energy constructions in different climate conditions opened following items for discussion:

- Lighting is the biggest energy consumer, only in the tropical environment cooling/drying of supply air is bigger; there are solutions for making lighting more energy efficient as seen in this comparison, but new solutions for supply air cooling/drying would be needed especially in tropical climate.

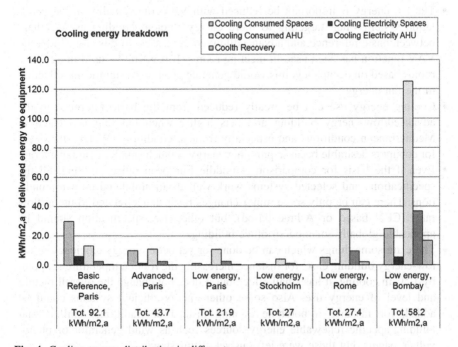

Fig. 4 Cooling energy distribution in different cases

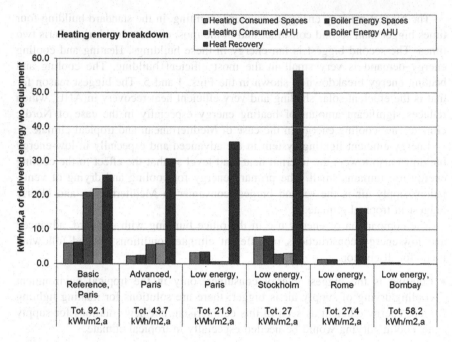

Fig. 5 Heating energy distribution in different cases

- The fan energy is important be reduced with VAV functionality in the venti-lation system when targeting to more energy-efficient building as seen here between basic reference and advanced/low-energy cases. In this case study, the VAV function has been used in meeting rooms. If it would be used in all office rooms based on occupancy, this would generate even more significant reduction in the fan energy.

- Cooling energy use can be greatly reduced from the basic reference to the advanced/low-energy building and gets higher when building is located in Mediterranean conditions and especially tropical conditions. Water–air system for cooing is desirable because pumping energy is much smaller than fan energy.

- Even if the basis for comparison is middle European office building, all the specifications and selected systems work well in all simulated climate condi-tions. There can be only some minor changes for instance related to the airflow rates (CEN based or Ashrae based), but otherwise, specification should be applicable globally for modern office buildings.

- There are some things which can be done for yet better energy efficiency in the low-energy building case, for instance increasing the room temperature set point by 1 °C in cooling, it has small effect, but it does not change the overall picture and level of energy use. Also some other energy-efficient systems could be added, for instance a borehole cooling instead of traditional chillers and building-specific renewable energy sources such as wind generator of photo-voltaic panels, but these were left out yet at this stage.

1.7 Concluding Remarks

The energy use of the office building with good standard construction, with construction including most common energy-efficient features, and with yet more modern technology for energy efficiency has been compared. The effects of different factors have been compared. Then, the most energy-efficient building has been analysed in different climate conditions: in Nordic and Mediterranean climates, and in tropical Asian climate. The energy use can be strongly reduced from basic reference building by using:

- energy-efficient structures especially for windows and solar shading.
- modern lighting system with intelligent controls.
- optimal HVAC system with very efficient heat recovery, good chiller design and air–water based chilled beam system with VAV functionality.

When building is located in Mediterranean or tropical climate conditions, big part of energy use comes from cooling/drying of supply air. All the specifications and selected systems fit in well to different climate conditions when designing modern energy-efficient office building.

2 Energy Targets in Architectural Competitions

Architectural competitions are one early-stage planning and design phase used typically for larger or more demanding or monumental buildings. The problem is that if energy performance targets will be applied after architectural competition (i.e. not included in the competition brief), this might be too late, and in worst case, the whole proposal has to be redesigned to meet the targets. This easily raises the question that the wrong entry has won the competition. To avoid such problems, energy performance targets are to be included in the competition brief among all other targets. The ultimate question is, how to do this in a proper way, so that:

- Energy targets will lead to integrated design and are considered from very first steps as massing and orientation issues;
- Unnecessarily complicated and detailed analyses can be avoided, because in very early stages, more robust and faster approaches are justified;
- All competition entries can be compared in fair way, i.e., everybody uses the same input data and reporting format, and energy performance is achieved with good design instead of input data manipulation.

In the following, energy targets and competition models are discussed based on experience from three recent international architectural competitions in Finland (Synergy, Low2No and Helsinki Central Library) and one smaller competition in Estonia.

2.1 Quantitative and Qualitative Targets

Energy targets are one issue in the sustainability, which can be measured with economic, environmental and social factors (EN 15643-1:2010). These categories could be measured with investment and life cycle cost, CO_2 emissions from energy and building material production and with indoor environmental quality (discussed in Chap. 5). It is important that quantitative performance indicators from all these three categories are included in competitions; however, they are not the most important ones. The main purpose of the architectural competitions is usually to find the best architectural and cityscape solution which has to come with excellent functionality and be as sustainable as possible. Therefore, in majority of competitions, sustainability and energy targets support the main targets. In technology-oriented or sustainable design development competitions, these targets can be also in the major role. In typical architectural competitions, the following categories of the assessment criteria are used:

1. Cityscape (compatibility with the site and fitting into the urban fabric);
2. Architecture (architectural design of the exterior and interior);
3. Usability (functionality/quality of working environment);
4. Ecological sustainability (indoor climate, energy performance and material efficiency);
5. Feasibility (construction and life cycle costs, possible to construct, operate and maintain).

When two last categories can be measured with quantitative (numeric) performance indicators, first three categories need qualitative assessment based very much on comparison of entries. Assessment of the competition entries is not a simple summing of scores of each category, because these categories had to sum up with sound overall solution and had to have good development potential— commonly required in competitions.

Quantitative nature of two last categories provides two options to specify assessment criteria:

- As minimum performance requirements, i.e., energy performance of X kWh/(m² a) primary energy has to be achieved, and for better performance, no credit is given;
- As a reference performance level which has to be achieved, but the entry with the best performance will receive the highest score, which is the typical assessment also for qualitative criteria (architecture, etc.).

In practice, there is no big difference which option is used, because if numeric performance indicators are required, they are taken into account by teams, and to do this, an integrated design approach is used that was the main purpose of such indicators. It is more important to define transparent and enough robust calculation procedure and input data for the calculation of performance indicators by teams. Some performance indicators could be better left to the jury, to be calculated during the assessment process of the entries. Typically, the construction cost has been calculated by the same consultant working for the jury, conducting the cost calculation of all entries.

2.2 Competition Models

Architectural competitions can be classified as one- or two-stage competition. One-stage competitions are with limited number of teams (qualified or invited) and energy analyses, and other calculations can be quite easily required. Two-stage competitions (especially international ones) may have many hundred up to about thousand entries in the first stage which means that energy calculations cannot be required during first stage and more simple criteria and verification has to be used. However, for the best proposals selected to the second stage, energy assessment is needed in order to be sure that the proposal could fit or could be developed to fit with energy performance and other numeric targets. In the second stage, similar calculations can be easily required as in one-stage competitions. To require the calculations, the calculation procedure, input data and reporting format have to be carefully specified in the brief.

2.3 Specification of Indoor Climate, Energy and Material Efficiency Targets

Indoor climate targets can be specified according to indoor climate classes of EN 15251 [5] discussed in Chap. 5 (or corresponding national code or standard). In the context of architectural competitions, it means a very short specification, including required room temperatures in winter and summer, ventilation rates and lighting levels. These values are needed also as input data, if energy simulations would be required.

Energy performance targets specification depends on assessment method used. There are two basic options:

1. To require energy simulations of a whole building and to specify energy performance target as primary energy;
2. To use simple indirect indicators and not to require energy simulations.

First option needs much more effort and also a very careful specification of the calculation procedure in the brief. In the case of two-stage competitions, energy simulations will be done in the second stage. This method was used in the case study reported in Sect. 3.

Second option does not enable the use of primary energy indicator, but more simple indirect energy performance indicators have to be used. Based on building envelope area data, the specific heat loss per room programme area can be very easily calculated as shown in Fig. 6. This simple indicator (with fixed building envelope element thermal properties) allows to control massing and façade design efficiency especially in heating-dominating climates, but is relevant for all European climates. To control cooling load and energy, very simple temperature simulations of some single typical rooms have to be required, and maximum cooling load target value has to be specified in W/m^2. This method (to fill in the table shown in Fig. 6 and temperature simulations of some representative rooms) can be seen as minimum for energy performance assessment. In two-stage competitions, the Table can be required in the first stage (and with final values in the second stage) whereas temperature simulations are relevant in the second stage. Main limitations of this method are cooling energy (cooling load provides some indications) and daylight which cannot be assessed. Heating energy cannot be directly seen as well, but as the specific heat loss coefficient correlates well with space heating energy need the entries can be compared adequately (the lower the specific heat loss, the lower the heating energy need).

In the case of both energy performance assessment methods, some graphical descriptions of HVAC and façade technical solutions are good to require in the brief for the assessment of entries. One schematic cross section of the building

Competition entry					
Room program floor area, m^2	1.0	Net floor area, m^2	1.0	Gross floor area, m^2	1.0
Heat losses through building envelope components				**Infiltration heat losses and thermal bridges**	
	U_i, $W/(m^2 \cdot K)$	A_i, m^2	H_{cond} W/K		
External wall	0.15	1.0	0.2	q_{50}, $m^3/(h\,m^2)$	1.5
Roof	0.09	1.0	0.1	No of storeys	1
External floor	0.12	1.0	0.1	q_{inf}, m^3/s	0.0000
Windows	0.80	1.0	0.8	H_{inf}, W/K	0.1
	A_{env}, m^2	4.0		Share of thermal bridges, %	20
	H_{cond}, W/K		1.2	H_{tb}, W/K	0.2
Specific heat loss of the building envelope, $H = H_{cond} + H_{tb} + H_{inf}$				H, W/K	1.4
Average U-value of the building envelope				H/A_{env} $W/(K\,m^2)$	0.4
Specific heat loss per room program floor area				$H/A_{Room\ program}$ $W/(K\,m^2)$	1.4

Fig. 6 Simple worksheet calculator for specific heat loss calculation with fixed (*grey shading*) values for all competition entries. *Yellow fields* are to be filled in—four building envelope area values are needed for calculation. Net floor area and gross floor area are additional information (for the efficiency assessment of entries) not used in the specific heat loss calculation

showing the operational concepts of the technical systems and façade solutions has been enough and has worked well in practice for this purpose. Such section should show ventilation, heating, cooling, daylight and solar shading solutions, as well as any other relevant active or passive solutions used. Mechanical room locations should also be shown and short explanatory text about technical concepts used has to be provided either in the same drawing or as an additional technical note of 1–2 pages.

If energy simulations will be required, it is important to fix main technical solutions in order to receive comparable results from all entries. This applies for building-site-dependent energy supply solutions (district heating, district cooling, gas, which renewable solutions can be used, etc.) which are to be defined. Similarly, the main parameters of ventilation (airflow rates, operation hours, heat recovery efficiency, specific fan power) are better to fix for energy calculation; however, other technical solutions for ventilation (mixed mode, another air distribution, etc.) could be accepted. If teams use other than the reference solution, they can assess energy savings with actual solution relative to the reference solution, that will make the assessment of results easier (instead of quite arbitrary results difficult to judge because of different solutions and system efficiency parameters used by teams, it can be seen how much savings have been accounted for each specific solution). For the cooling, it is at least relevant to define in which rooms a room conditioning has to be used (in addition to central cooling of supply air in AHUs).

All input data needed in energy calculations have also to be defined— occupancy schedules, internal heat gains (lighting, appliances, occupants), ventilation airflow rates, temperature set points in winter and summer, etc. Depending on the purpose, the U-values of the building envelope components could be fixed or not. Such limitations indeed reduce the freedom of design and should be well justified. Experience from competitions has shown that if the main technical solutions were not fixed, the entries ended up with solutions with highly inconsistent efficiency, ambition and cost, and the results were very difficult to compare without recalculation. If in addition to main technical solutions, the calculation procedure, input data and reporting format were well specified; for majority of the entries, the energy simulation results were assessed as reliable, and some recalculations were needed only in specific cases. Energy calculation procedure and input data have typically to follow national building code (relevant parts can be translated) and used as appendixes of the brief. If the building code does not support energy simulation, the energy calculation methodology has to be described in the brief that needs a significant effort; however, the general calculation principles are well known.

Material efficiency targets can be specified in similar fashion to primary energy. The specific CO_2 emission indicator shows how many kilograms of CO_2-emissions per floor area are released during the production of construction materials of main structures. Similarly to energy calculation procedure, the calculation of building material volumes has to be well specified. Such calculation is typically limited to the load-bearing structures and building envelope (finishing materials, partitions and other less important components will not be calculated). The calculation method is specified in EN 15978 [6] and a case study example is reported in Sect. 3.

3 Architectural Competition Case Study: Synergy in Helsinki

Viikki Synergy competition was one-stage competition held in 2010–2011, where six qualified teams prepared comprehensive design of about 20,000-m^2 office and laboratory building as shown in Fig. 7 [8]. The competition entries required were relatively detailed for such competition, including energy simulations and embodied carbon analyses. In this chapter, the assessment criteria for sustainability from quantitative measuring point of view are discussed.

The innovation of the competition was the assessment criteria for sustainability, summing up the energy performance and material efficiency data in kgCO$_2$/m^2 units in the assessment process. This criteria and lessons learnt from the competition can be utilized in future competitions in order to design and build sustainable buildings.

3.1 Assessment Criteria of the Brief

The competition brief used well-specified assessment criteria, from which roughly 50 % was quantitative (measurable with performance indicators as tons of CO$_2$ or Euros) and another 50 % qualitative ones related to architectural components. In Viikki Synergy, four main categories with roughly equal importance were as follows:

Ecological sustainability including energy performance and material efficiency

- Urban and architectural quality.
- Usability (functionality/quality of working environment).
- Feasibility (economic efficiency and quality of technical solutions).

Fig. 7 First-Prize-awarded-entry Apila of the competition (a low-rise large building in *front right*)

These categories had to sum up with sound overall solution and had to have good development potential. Referring to good architecture, reasonable cost and sustainable use of energy and material resources, the categories were supported by transparent assessment framework well described in the brief.

Ecological sustainability was measured with energy performance and material efficiency. Energy performance followed the target of EPBD recast for 2019–2021, nearly zero-energy buildings, which was the basis for energy performance target value of 80-kWh/(m^2 a) primary energy without tenants electricity (all other energy flows included according to EN 15603). It was assessed that 80 kWh/(m^2 a) per programme area will correspond roughly to 70 kWh/(m^2 a) per net area (the difference is caused by corridors not included in the room programme). Energy carrier factors to calculate the target of 80 kWh/(m^2 a) were 2.0 for electricity, 0.7 for district heat and 0.5 for renewable fuels. For the energy supply systems, it was specified to use on-site solar electricity production corresponding to 15 % of total electricity use (facility + tenant electricity). This fixed amount was justified with high cost of PV-panels, and making it easier to compare the proposals. All other solutions for energy performance were let open.

Comprehensive energy performance calculation guidance was provided as the appendix of the competition brief. This was necessary, because the primary energy calculation frame provided in the Finnish building code D3 2012 was not available. In future competition briefs, this part can be simply replaced by the reference to relevant calculation frame, such is the building code in the Finnish case.

Material efficiency was measured in kgCO$_2$/m^2 floor area and teams competed to achieve as low value as possible without compromising with other criteria. The assessment was limited to the main structure's carbon footprint that was derived from the carbon dioxide emissions resulting from the building materials' manufacture and the materials' possible carbon dioxide storage. For the material emission calculation, the specific emission values were provided in the brief as shown in Table 3.

In the assessment, the energy performance and material efficiency data were summed in kgCO$_2$/m^2 units by the use of specific emission factors for energy carriers instead of primary energy factors. Such assessment resulted in life cycle CO$_2$ emissions, as well as LCC in the economic efficiency assessment. For the LCC, the jury ordered construction cost calculations from the consultant not being involved in the completion (i.e. cost calculations were not included in the brief). The same consultant provided cost calculations for the all six proposals.

Energy performance was also recalculated by another consultant for two proposals. As the results were very close to those provided by the competition teams, the energy calculation of the rest of proposals were not recalculated to save time and money.

Relatively easy and fast cost and energy calculations as a part of the assessment procedure were possible, thanks for the building information models required in the brief. These BIM models made it possible to analyse the proposals with the software tools used for cost and energy calculations.

As a result of assessment, the proposals were compared in the life cycle carbon (tons of CO$_2$) and life cycle cost (M€) scale, Fig. 8.

Table 3 Specific emission factors for the material emission calculation. Due to somewhat local origin, the values have mainly Finnish references and apply in Finnish conditions. Such data can be found from environmental product declarations according to EN 15804 [7]

Material	Mass per volume kg/m³	Greenhouse gas emission gCO$_2$-eq/kg[a]	Carbon storage g CO$_2$-eq/kg	Source[c] Appendix 19.2
Concrete products				
Steel used for concrete reinforcement	7,850	440		IISI European average, EAF-route
K35 ready-mixed concrete	2,400	140		RT environmental declaration
K90 ready-mixed concrete	2,400	200		RT environmental declaration
K60 multicore slab	2,400	170		RT environmental declaration
K60 double T-slab	2,400	210		RT environmental declaration
Metal products				
Steel pipe	7,950	1,090		RT environmental declaration
Structural hollow sections, steel pipe pics, steel sections	7,850	1,090		RT environmental declaration
Welded steel beam	7,850	780		Rautaruukki
Hot-dip galvanized steel products	7,850	1,040		Rautaruukki
Cold-rolled steel sheets and coils	7,850	880		Rautaruukki oyj IISI-data Ruukki
Hot-worked steel sheet	7,850	730		Ruukki, RT environmental declaration
Paint-coated steel sheets and coils	7,850	1,070		RT environmental declaration
Steel wire rope	7,300	2,680		Fundia wire oy
Aluminium profile (85 %)	2,700	3,640		Scanaluminium
Brickwork products				
Lime brick for external cladding	1 750	180		Optiroc Oy
Lime brick for dividing walls	1,750	150		Optiroc Oy
Burnt clay brick	1,300	220		Wienerberger Ov
Block of lightweight aggregate concrete	700	330		Betonikeskus ry

(continued)

Table 3 (continued)

Material	Mass per volume kg/m³	Greenhouse gas emission gCO$_2$-eq/kg[a]	Carbon storage g CO$_2$-eq/kg	Source[c]
Expanded clay aggregate concrete pand	880	270		Rakennusbetoni- ja elementti Oy, Appendix 19.2
Autoclaved aerated concrete	440	280		RT environmental declaration
Building mortar M100/600	1,800	140		Optiroc-house
Block mortar M100/500	1,800	220		Optiroc-house
Thin mortar joint	1,800	220		Optiroc-house
Lime plaster 50/50	1,800	130		Optiroc-house
Average filter	1,800	170		VTT research report
Concrete roof tile	44	140		Ormax, lafarge roofing oy
Wood and board products				
Glued laminated timber	440	330	1,600	RT environmental declaration
Sawn umber	480	70	1,600	RT environmental declaration, VTT optiroc-house
Dried floor board/plank	480	90	1,600	LCA-saha data-sheet
Planed board/plank (for indoor use)	480	110	1,600	LCA-saha data-sheet
Glued laminated bean	500	230	1,600	Pöyry
Fibreboard, softboard	300	420	1,600	RT environmental declaration
Medium density fibreboard	780	270	1,600	Karlit ab/Trätek envi. decl.
Hardboard	950	130	1,600	Karlet ab/Tratek envi decl.
Chipboard	670	390	1,600	VTT research report/Puhos-board Oy
Standard conifer plywood	450	680	1,600	RT enviroment declaration
Standard birch plywood	660	720	1,600	Puuinfo Oy, RT environmental declaration
Gypsum board	9 or 11.7	390		RT environmental declaration
Plastic products				
Polyethene (HD)	940	2,410		Association of Plastics Manufacturers in Europe
Polypropylene	905	2,100		ELCD database, polypropylene fibres (PP)

(continued)

Table 3 (continued)

Material	Mass per volume kg/m³	Greenhouse gas emission gCO$_2$-eq/kg[a]	Carbon storage g CO$_2$-eq/kg	Source[c]
LDPE (Low-density polyethylene) foil	940	2,100		Apme.org
Insulating materials				
Wood fibre insulant from collected paper—Termex	26–65	220	800	RT environmental declaration
Wood fibre insulant from collected paper—Ekovilla	30–65	180	800	RT environmental declaration
Expanded clay aggregate/lightweight aggregate	290	420		Optiroc Oy
Stone wool	22–250	990		RT environmental declaration
Glass wool	20	800		Isover Oy
PUR. polyurethane		4,230		VTT calculations on SPU-insulant
Polystyrene, floor board	15	3,600		Apme org
Extrusion-compressed polystyrene	38	3,600		Apme org
Other materials				
Bitumen–polymer sheeting	–	1,120		VTT optiroc-house
Sheet glass flat glass	2,500	BCD		VTT energy-corrected Pilkington Ltd
Window frame	Calculate yourself in the Excel sheet using the material amounts			
Reused (i.e. old ones) structures and materials[b]	0			

[a] At factor gate, the jury may complement the calculation by the emissions of the transport (especially its long distance)

[b] The jury may complement the calculation lay emissions from the processing and transport of the material

[c] A large part of the sources and values collected by Timo Rintala/Pöyry RT Environmental declarations: www.rts.fi

Rough estimates of the carbon storage were based on the RT Environmental Declarations of following products: sawn timber, glued laminated timber, birch plywood. The carbon storage of wood fibre insulants, made from waste paper, is a very rough estimate based on materials of the insulant.

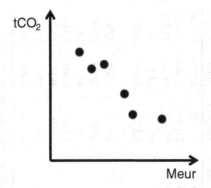

Fig. 8 Ecological and economical performance map used in the assessment of proposals in Viikki Synergy competition. *All dots* (six proposals) are fictive examples, representing the jury expectations about less ecological proposals with lower cost (*dots in the left*) and more ecological proposals with higher cost (*dots in the right*). See the real results reported in next section

3.2 Results

The key figures for the competition entries' energy and materials efficiencies are shown in Table 4. Primary energy values describing total energy use calculated with energy carrier factors (2.0 for electricity and 0.7 for district heating) are first presented in MWh/a units for a reference building solution complying with valid minimum code requirements, and for the design solution with conventional energy supply solutions as specified in the competition programme. Primary energy for the actual design solution is then presented in MWh/a and kWh/(m^2, a) units, of which the latter reflects the programme floor area, and does not include user electricity according to the competition programme's definition. The energy use of the actual design solution has been calculated as the CO_2-e emissions caused by 30 years of energy use, with an approximate specific emissions factor of 150 kg CO_2-e/MWh for next 30 years. This specific emissions value was used both for electricity and district heating.

The main structure's carbon footprint was derived from the carbon dioxide emissions resulting from the building materials' manufacture and the materials' possible carbon dioxide storage. Solaris has functioned as a carbon sink because its carbon dioxide storage has been larger than the emission caused by the manufacture of its building materials. The breakdown of the carbon footprint is shown in Table 5.

The table's bottom line shows the sum of 30-year energy use and the main construction's carbon dioxide emissions. This key figure serves as the estimate for the property's 30-year carbon footprint.

The results show that the leaders are Apila in energy performance and Solaris in material efficiency. The results for energy performance are fairly even, with Pastorale, however, somewhat separated from the rest. In terms of material efficiency, 191910 and Pastorale are clearly weaker than the other entries.

Table 4 Key figures for competition entries' scope and costs, as well as energy and materials efficiencies; the three best results are shown in boldface

Competition entry	1 Solaris	2 Valaistus	3 Pikkukampus	4 Pastorale	5 Apila	6 191910
Rentable area, m² (programme area 12,855 m²)	14,000	13,100	14,600	15,800	12,700	14,800
Net floor area, m²	18,300	18,800	18,200	21,500	19,800	20,400
Gross floor area, m²	20,600	20,300	20,100	23,600	21,800	23,800
Construction cost estimate, M€	**54.9**	**54.6**	58.5	57.7	**54.1**	61.1
Primary energy, reference building solution complying to minimum code requirements, MWh/a	5,104	**4,513**	4,575	**4,272**	4,523	**4,053**
Primary energy, design solution with conventional energy supply solutions, MWh/a	3,576	3,517	**3,502**	3,631	**3,508**	**3,281**
Primary energy, actual design solution, MWh/a	2,851	**2,765**	2,830	2,985	**2,674**	**2,780**
Primary energy, actual design solution, kWh/(m² a) without user electricity	99	**92**	97	109	**85**	**93**
30-year carbon dioxide emissions of energy use, tCO₂-e/a	7589	7146	**6,904**	7,726	**6,005**	**6,102**
Carbon footprint of main structure, tCO₂-e	**−470**	**147**	1,600	3,269	**481**	4,013
Carbon footprint of main structure, kgCO₂-e/m²	**−37**	**11**	124	254	**37**	312
Total carbon footprint of 30-year energy use and main structure, tCO₂-e	**7,119**	**7,293**	8,504	10,995	**6,486**	10,115

The key figures for energy use have been checked so that energy use categories reported are identical in all entries. Additionally, to check the results, energy simulations were carried out for two entries, whose results were very close to the figures provided

Table 5 Material emissions and carbon storage of competition entries

Competition entry ($kgCO_2$-e/m^2)	1	2	3	4	5	6
	Solaris	Valaistus	Pikkukampus	Pastorale	Apila	191910
Material emissions of main structure	179	151	n.a.	256	260	646
Carbon storage of main structure	−215	−140	n.a.	−2	−222	−335
Carbon footprint of main structure	−37	11	124	254	37	312

Carbon footprint used in the assessment was calculated as a sum of material emissions and carbon storage. Note that the calculation method of EN 15978 [6] not taking carbon storage into account in the life cycle was not available at the competition time.

When assessing the 30-year total emissions, Apila's 6,500-t emissions are the lowest. Next are Solaris and Valaistus at 7,100 and 7,300 t, respectively. Pikkukampus is situated midway on the scale at 8,500 t, and the two remaining competition entries 191910 and Pastorale are clearly weaker than the others, exceeding the 10,000-t limit.

In the main structure's carbon footprint calculation, both material emissions and carbon storage were taken into account. Full inclusion of the carbon storage is a simplification that affects remarkable results and means the assumption that materials will stay forever in the building structures. Another and more accurate possibility will be to include the carbon storage only for materials reused or recycled after demolishing of the building. The effect of carbon storage is shown in Table 5. Pastorale had mainly concrete structures, and the carbon footprint is much higher compared to mainly wooden entries Solaris, Valaistus and Apila. In material emissions, the difference is less significant, showing the meaning of the carbon storage treatment in the calculations.

The results of the competition works' ecological sustainability were compared to life cycle costs as certain competition entries have required more substantial actions to achieve good energy performance, which for their part affects construction costs. An extreme example of this was 191910, in which the building's uneconomical shape (a remarkably larger external surface area than the other entries) has been compensated with a clearly enhanced level of thermal insulation, as well as with a low-pressure ventilation system integrated with a structure. Thirty-year life cycle costs (calculated as the sum of estimated construction costs and estimated 30-year energy expenses) are compared to 30-year total emissions in Fig. 9.

From the standpoint of ecological sustainability, a cost comparison of the three best competition works (Apila, Solaris and Valaistus) demonstrated, within the framework of calculation accuracy, virtually identical construction costs, the cost difference for these three entries falling within a range of 1.5 %. The values of the estimated construction costs for Pastorale, Pikkukampus and 191910 were significantly larger (+7–13 % compared to the most economic one). Thus, the best in terms of energy performance, Apila, was also the best in terms of life cycle costs, with Valaistus and Solaris following close behind.

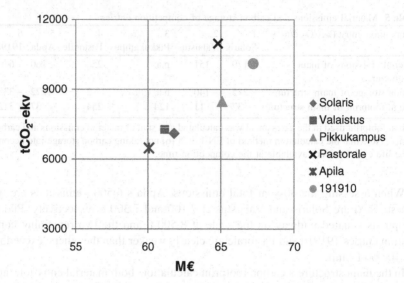

Fig. 9 Placement of competition entries on scale of 30-year life cycle costs and carbon footprint of main constructions and 30-year energy use. The competition entries form two fairly distinct groups; the group formed by Apila, Solaris, and Valaistus simultaneously has clearly lower emissions and costs

3.3 Conclusions

Competition entries have shown that there are not necessarily conflicts between sustainability and architectural categories; as in many cases, these different categories can support each other and lead to proposals with different and rich architecture.

Inclusion of quantitative energy performance, material efficiency and economic efficiency targets will direct the design and selected concepts from very first steps of design teams. Design teams have shortly noticed that integrated design is needed to meet the performance criteria. However, bearing these criteria in mind, there is still a lot of room for functionality and architectural components. The criteria used did not limit the architectural quality that was demonstrated by various massing alternatives proposed.

It may also be seen so that if all teams meet exactly the specified quantitative performance targets (kWh/m^2, kgCO$_2$/m^2 at roughly the same cost), the winner will be selected very much based on functionality and architectural components, which is not different from traditional architectural competitions. In such a case, quantitative performance targets just have assured the technical quality of the proposals, i.e., being energy, material and cost efficient. In reality, there are usually differences between the proposals, i.e., how well they meet energy and material targets and quite often in the economic efficiency, and these differences may serve as decision bases between proposals with roughly equal architectural quality.

Best proposals of Viikki Synergy competition were able to sum up high architectural quality, good functionality, energy and material efficiency as well as cost efficiency. It can be concluded that for the jury, there was no need to select between ecologically efficient and cost-efficient entries, but selection was made within the group of entries being both ecologically and cost efficient.

For the future sustainable design competitions, there are some issues in the competition programme to be further developed. Primary energy calculation can be done in Finland according to new building regulation, Finnish building code, Part D3 2012, which was not available at the competition time. D3 2012 specifies similar calculation framework as was in the annex of the competition programme; therefore, this annex can be simplified. Carbon footprint calculation was not fully standardized at the competition time and will still need detailed guidelines in order to achieve meaningful comparable results. Inclusion of the carbon storage in the building life cycle assessment was not correct according to EN 15978 [6] calculation method, which specifies carbon storage assessment in the supplementary information module beyond the building life cycle, dealing with materials reuse and recycling. Therefore, it can be recommended to limit carbon assessment in architectural competitions to building life cycle, meaning that carbon storage assessment would not be done. Another detail in carbon calculations were the foundations causing some confusion. In order to keep reasonable accuracy, it can be suggested to provide model solutions for foundations in the competition programme so that reasonable alternatives for lightweight and heavyweight structures and construction frame types are available with load-bearing capacity and carbon footprint data. This will avoid the unnecessary effort of foundation sizing as well as possible under or over sizing of foundations what was suspected in this competition in a couple of cases.

References

1. RT (2005) Réglementation Thermique 2005: Des bâtiments confortables et performants. Centre Scientifique et Technique du Bâtiment, France
2. Loutzenhiser P, Heinrich M (2007) International energy agency's Task 34/Annex 43 Project C report. Empirical validation of shading/daylight/load interactions in building simulation tools. International energy agency
3. Kurnitski J (2009) Role of ventilation and cooling in the energy balance of modern office buildings. REHVA European HVAC J 46(4):39–44
4. Virta M, Butler D, Gräslund J, Hogeling J, Lund K, Reinikainen E, Svensson G (2004) Chilled beam application guidebook. REHVA Guidebook no. 5, REHVA 2004
5. EN 15251 (2007) Indoor environmental parameters for design and assessment of energy performance of buildings addressing indoor air quality, thermal environment, lighting and acoustics. European committee for standardization, CEN 2007

6. EN 15978 (2011) Sustainability of construction works—assessment of environmental performance of buildings—calculation method. CEN 2011
7. EN 15804 (2011) Sustainability of construction works—environmental product declarations—core rules for the product category of construction products. CEN 2011
8. Kurnitski J (2011) Lessons learnt from Viikki Synergy building sustainable development design competition: proposed criteria for sustainability. SB11, world sustainable building conference, Helsinki, Finland, 18–21 Oct 2011

Basic Design Principles of nZEB Buildings in Scoping and Conceptual Design

Hendrik Voll, Risto Kosonen and Jarek Kurnitski

Abstract nZEB buildings generally require integrated design in order to achieve design targets economically. Decisions and choices in early design stages may be expensive or even impossible to fix later if have not been successful. Massing not supporting energy-efficient design or lack of space for technical systems is typical example of potential drawbacks. It is important continuously to follow that design targets can be met. In early stages, rules of thumb and some key parameters can be used for indirect assessment, which is the method until first energy simulations can be run. Next step is to be sure that planned technical systems can be fitted in the building—there has been enough mechanical space and proper locations enabling energy-efficient design. These and other important milestones in the early stage including fenestration design, shadings and daylight are discussed in this chapter. It is not enough to design a good nZEB building, but it has to be done in a way that the building can be also operated as nZEB building. In majority of projects, designed room layouts will change already during construction, because of clients' needs. Therefore, the HVAC systems must adapt to changed loads and partition wall locations. To enable flexible space use and adaptive systems, special considerations and the use of room modules are needed, that is, the last but not least issue discussed in this chapter.

1 Key Parameters for Early-Stage Energy Performance Assessment Without Simulations

Assessment of energy performance, are the targets met or not, is most difficult in early stages of the design. Without drawings, nothing can be calculated. To follow how energy performance will build up can be done through energy balance

H. Voll (✉) · R. Kosonen · J. Kurnitski
Tallinn University of Technology, Ehitajate tee 5, 19086, Tallinn, Estonia
e-mail: hendrik.voll@ttu.ee

J. Kurnitski (ed.), *Cost Optimal and Nearly Zero-Energy Buildings (nZEB)*,   103
Green Energy and Technology, DOI: 10.1007/978-1-4471-5610-9_7,
© Springer-Verlag London 2013

components and decisions of basic technical solutions. The pyramid in Fig. 1 shows the correct order of choices in the design process and the impact of those on energy performance and cost. This approach together with follow-up of limit values of energy performance key parameters is the only way to control and plan energy performance of a building in early stages until first energy simulations can be done for the sketch variants. Simulation will show the magnitude of energy balance components, which allows us to continue more detailed work under each category.

Massing and orientation on the site have crucial effect on compactness and energy performance. The control of external building envelope and the size of glass facades can improve energy performance so much that it could not be compensated in next steps of the pyramid if failed. Well-known massing measures especially for a cold climate are the use of courtyards as covered atriums to improve compactness and double-skin facades to reduce window area (but still maintain the look of "glass" building).

If the shape and fabric have somehow formed, it can be continued with façade design and selection of technical systems. Facades need to provide simultaneously views, daylight, solar protection and thermal insulation. From technical systems, the ventilation system (often combined with air conditioning) has the largest space requirements. Energy-efficient, dedicated outdoor air systems need larger air-handling units (AHUs) and ductworks compared to conventional mechanical or mixed-mode systems. Mechanical rooms, risers and air intakes need to be effectively fitted into architecture of the building.

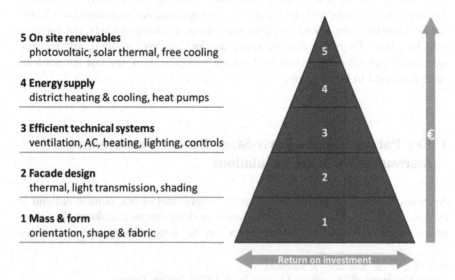

5 On site renewables
photovoltaic, solar thermal, free cooling

4 Energy supply
district heating & cooling, heat pumps

3 Efficient technical systems
ventilation, AC, heating, lighting, controls

2 Facade design
thermal, light transmission, shading

1 Mass & form
orientation, shape & fabric

Return on investment

Fig. 1 Energy hierarchy has similarity with construction works: without foundations, it is not possible to raise next floor. Energy performance-related choices are reasonable to do in logical order

Technical systems are combined with energy supply solutions, which in the simplest cases can be district heating and cooling. nZEB buildings compensate energy use with on-site renewable energy production, which are most often solar systems.

The arrows of the cost and return of investment stress the importance of the choices done in the bottom steps of the pyramid. In these steps, the energy performance measures are relatively low cost and have bigger impact compared to more expensive measures with limited impact at higher steps of the pyramid. For example, mistakes in massing cannot be compensated with on-site renewable energy. As a rule of thumb, all steps of pyramid need to be carefully solved in nZEB buildings.

The limit values of energy performance key parameters can be used in early stages to assess the compliance with energy performance target. Such values are provided in Table 1 for Northern European climate. The only difference between low-energy and nZEB building in this table is in on-site renewable energy production (this follows Estonian approach according to which on-site RES is not necessary in low-energy buildings). All other values are the same for low-energy and nZEB buildings. BAU values correspond to good construction practice.

Table 1 Limit values of energy performance key parameters for low and nZEB buildings in Northern European climate

	BAU	Low-energy and nZEB
Specific heat loss of building H/A_{floor}, W/(K m^2)		
$A_{\text{floor}} = 500$ m^2	0.8	0.4
$A_{\text{floor}} = 1,000$ m^2	0.6	0.3
$A_{\text{floor}} = 2,000$ m^2	0.5	0.25
$A_{\text{floor}} \geq 4,000$ m^2	0.4	0.2
Building envelope average U-value, H/A_{env}, W/(K m^2)	0.5	0.25
Windows, average U-value, W/(m^2 K)	1.4	$\leq$0.7
Windows, solar factor g, (–)	0.3–0.5	Optimized
External walls, U-value, W/(m^2 K)	0.25	0.14–0.18
Building leakage q_{50}, m^3/(h m^2)	$\leq$3	$\leq$0.6
Solar shading		External
Average daylight factor, %	2	2
Window-to-wall ratio, %	40–90	25–30
Cooling load in typical rooms, W/m^2	50–100	$\leq$40
Heat recovery temperature efficiency, %	$\geq$70	$\geq$80
Specific fan power of ventilation SFP, kW/(m^3 s)	2–2.5	1–1.5
Demand-controlled ventilation		Meeting rooms and large spaces
Seasonal energy efficiency ratio of cooling ESEER	2–3	$\geq$5
Installed lighting power, W/m^2	$\leq$12	$\leq$5
Lighting control	Time	Dimmable lights and multi-sensors
Primary energy efficiency of heating		$\geq$0.90
Share of renewable energy, %		nZEB $\geq$ 10

The limit values are indicative, intended for the use in early design stages, and the final values will be determined by energy simulations

Specific heat loss coefficient H/A_{floor} calculated per net floor area of the building is used to describe heat losses of the building. This specific heat loss includes all impacts of compactness, size of windows and the level of thermal insulation. Specific heat loss is calculated as sum for all external building envelope elements with Eq. 1 in Chap. 4. For a building element, it is a product of the U-value and area. Thermal bridges and infiltration losses are also included in the specific heat loss. Because the compactness of the building (A_{env}/A_{floor} or A_{env}/V where A_{env} is the surface area of the building envelope and V is the volume of the building) will significantly improve with the size of the building, the table provides values depending on the size of the building. Thermal transmittance values of the building envelope and its components are all subvalues that include major specific heat loss coefficient H/A_{floor}. The average thermal transmittance of the building envelope equals to H/A_{env}, where A_{env} is the surface area of the building envelope. Good compactness and controlled window-to-wall ratio allow somewhat to exceed these limit values, but still comply with the limit of the specific heat loss coefficient H/A_{floor}.

Window-to-wall ratio, solar factor g, solar shading and cooling load form one group of parameters, which affect daylight and cooling needed. Window size and visible light transmission have to be selected so that minimum average daylight factor is achieved in daylight zone where typically workplaces are located. At the same time, to minimize primary energy (or energy cost) from heating, cooling and electric lighting, the facades need to be optimized regarding these parameters. Fenestration design principles are discussed in Sect. 3.

External solar shading allows us to use clear glazing units with high light transmittance, which allows us to reduce window size without compromising the daylight factor and results in significant savings in heating and cooling energy (applies especially in a cold climate). "Glass buildings" are problematic in a cold climate because of heat losses, but the look of glass buildings can be achieved with double-skin facades. Double-skin facades allow easy and protected installation of external shading between the skins.

Cooling load can be easily simulated with single-zone models of critical rooms. Typically, one room module on south and west facade and one corner room are to be simulated.

Parameters related to ventilation require enough large mechanical rooms, risers and floor height for low-pressure ductwork, discussed in Sect. 2. Because of these space requirements, the technical concept of ventilation (and air conditioning) needs to be decided already in the scoping, in order to fit ventilation system properly to building.

Basic solutions of electric lighting are to be decided in conceptual design, when it is necessary to decide should the lights integrated in room conditioning units (in case of chilled beams, ceiling panels) or not. Both options allow us to use effective lights and combine direct and indirect lighting. Indirect lighting provides the best visual comfort but increases electricity use.

Room conditioning solutions have to be selected in conceptual design. Modern room conditioning units, chilled beams and ceiling panels, allow us to use energy-efficient high-temperature cooling and utilization of free cooling. A key solution for achieving small cooling energy use is the minimized cooling need complemented with effective chiller.

Primary energy efficiency of the heating system is defined as a product of the distribution and emission efficiency and generation (boiler) efficiency, what is divided with primary energy factor of heating energy carrier. If the efficiency of radiator heating is 0.95 and of a condensing gas boiler is 0.95, the primary energy efficiency of the heating system is $0.95 \times 0.95/1.0 = 0.90$, where 1.0 is the primary energy factor of gas. In the case of a heat pump, instead of efficiency of the generation, the seasonal performance factor of a heat pump will be used. For the heat pump with seasonal performance factor of 3.5, the primary energy efficiency of the heating system is $0.95 \times 3.5/2.5 = 1.33$, where 2.5 is the primary energy factor of electricity.

Easiest ways to produce on-site renewable energy are solar collectors (for water heating), heat pumps and solar photovoltaic cells (for generating electricity). In non-residential buildings, where the use of domestic hot water is low, it is difficult to apply solar thermal energy, because there is no heating need during solar thermal production. Therefore, the most common choice of solar energy is PV cells.

The values in Table 1 should not be taken as final truth, but their aim is to direct the choices and decisions in the early stage so that when energy simulation can be done, the result would fulfil low-energy or nZEB primary energy target. First energy simulation is appropriate to done within the scoping. To see proper basic solutions, a simple energy simulation model of one typical floor is often enough. Such model is also suitable for fenestration analyses in conceptual design in order to find the most rationale and economic facade solution for set energy performance target. In the calculation of primary energy target with the model of one typical floor, a safety margin of about 15–20 % should be applied, because external roof and floor are not considered, and typically, the entrance floors have larger glass surfaces, overhangs, larger airflow rates, etc., factors that are increasing energy use. In fenestration analyses, daylight simulations are also needed to support energy simulations in order to achieve even distribution of daylight and avoid glare.

2 Space Requirements and the Location of Mechanical Rooms

In order to achieve performance criteria of HVAC systems given in Table 1, mechanical rooms should be enough large and have appropriate shape and location. Especially, the heat recovery and specific fan power (SFP) of ventilation depend directly on the space need. The smaller the air-handling AHUs and ductworks, the worse the heat recovery and the higher the electricity use for transport of air. To achieve effective ventilation and cooling, the location of mechanical rooms has to allow installing simple, relatively short ductworks and cooling networks with low losses.

Mechanical rooms should be placed in the massing together with other spaces of the room program. Experienced architects and engineers study the space need and locate mechanical rooms already in first sketches. It is well known that later it could be impossible to find space or increase floor height. Packing HVAC into too tight rooms and the use of small ductwork may drop energy performance by one class of energy performance certificate, which may mean that low-energy building cannot be constructed. Additional consequences may be noise and balancing problems, which may lead to serious indoor climate deterioration.

The size (floor area and height) of mechanical room of AHUs depends on the total airflow rate. Total airflow rate can be estimated as a product of airflow rates given in Sect. 1 in Chap. 5 and net floor area. For example, in office buildings, an average airflow rate is 2 l/(s m^2) that equals 0.002 m^3/(s m^2) cubic metres per net floor area. The floor area and height of mechanical room are shown in Figs. 2 and 3. If energy supply equipment (district heating substation, compressors of air conditioning) is located in the same mechanical room, this will somewhat increase the required floor area.

In the mechanical room with optimal shape and location, AHUs can be placed near the air intake chamber, exhaust air can be directed through the ceiling to outdoors, and supply and extract ducts can be easily turned to the riser, which is shown in Fig. 4. Air intake is recommended from north or east façade, and exhaust air discharge should not be too close for air intakes or occupied zones. If outdoor or exhaust air has to be ducted through the risers (e.g., from the basement), this will increase the size of risers and the height of mechanical rooms. Optimal location of mechanical rooms will avoid excessively long ductworks. For this purpose, also the capacity of air-handling unit (AHU) should normally not exceed 6 m^3/s. In buildings not higher than 10 floors, the best location of AHUs is on the top of the roof or on the last floor. High-rise buildings have to be zoned with mechanical floors. Placing AHUs in the basement has to be avoided as the most ineffective regarding the space use and efficiency of ventilation system, because of very long outdoor and exhaust air ducting.

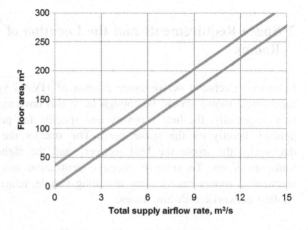

Fig. 2 Required floor area of plant rooms for AHUs as a function of total airflow rate. The size is to be selected according to the *upper curve*. If there is not enough space available, the risk can be taken and the *lower curve* can be used. The use of *lower curve* has to be checked by designing with specified air-handling units

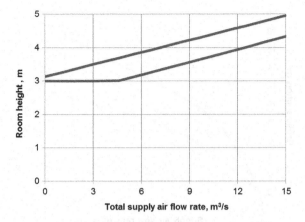

Fig. 3 Required height of plant rooms for air-handling units as a function of total airflow rate. The height is to be selected according to the *upper curve*. If there is not enough space available, the risk can be taken and the *lower curve* can be used. The use of *lower curve* has to be checked by designing with specified air-handling units

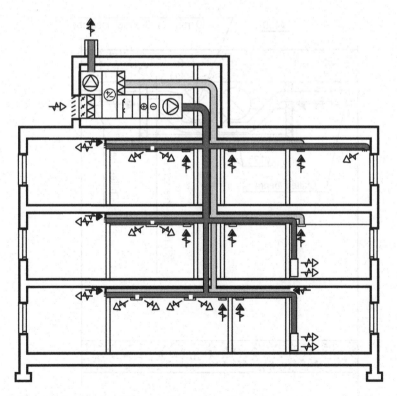

Fig. 4 An example of optimal location of an air-handling unit

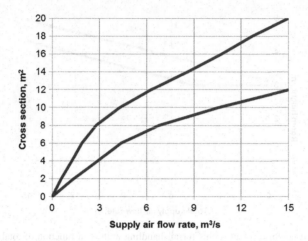

Fig. 5 Space need of risers

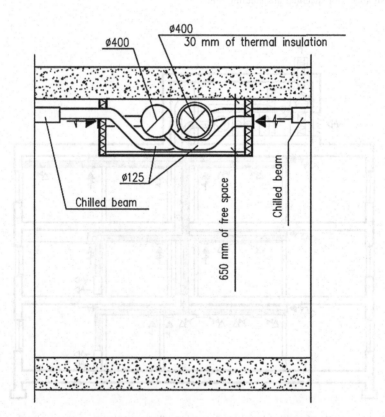

Fig. 6 An example of ventilation ductwork in the corridor or in the middle zone

Space requirement of risers depends on the airflow rate of ducts, other pipework installed as well as the location of mechanical rooms in the building. Room groups with similar use and operation time (offices, restaurant, toilets, etc.) will be typically equipped with own ventilation systems for flexible operation. If rooms are grouped and AHus are located so that serving the room groups will need many parallel ductworks, this will also increase the space need for risers. Space need of risers is shown in Fig. 5, but compared to the size of mechanical rooms, the spread is larger depending on the solutions in the specific building.

Enough large main ducts are needed for a low-pressure ductwork that enables flexible changes in the room program. If the room layouts are changed with consequent changes in ventilation needs, the low-pressure system can be relatively easily balanced to the new situation. This is a typical situation in operation; often, first layout changes are done already during the construction phase. Ductworks are typically installed in corridors or inner zones with suspended ceiling. Critical intersections of ducts have to be taken into account when specifying the floor height. Air distribution from main ducts can be done with standard solutions, and an example is shown in Fig. 6.

3 Fenestration Design Principles Based on Daylight and Energy Performance

3.1 Daylight Factor and Autonomy

Daylight is the combination of direct sunlight and diffuse daylight. Direct sunlight is the visible part of solar radiation with a clear direction. Diffuse daylight is the visible part of diffuse radiation, without a clear direction. To simplify, it can be said that direct solar radiation creates shadows of objects and diffused radiation does not.

Similarly, requirements for daylight are split into two categories:

- diffused daylight requirements are defined by a daylight factor;
- direct solar radiation requirements are defined in several European countries by solar insulation and its duration. However, in many European countries, there are no requirements for direct solar radiation.

The daylight factor is a ratio of internal illuminance to external horizontal illuminance:

$$\bar{D} = \frac{T \times Aw \times \Theta \times m}{A \times (1 - R^2)} \tag{1}$$

where
$\bar{D}$ is the mean daylight factor,

T is the visible light transmittance of the glass,
Θ is the angle of visible sky (in degrees),
m is the window glass soiling effect,
A is the total area of ceiling, floor and walls (including windows),
R is the weighted average of reflection factors of internal surfaces.

In calculating the daylight factor, it is assumed that the impact of direct solar radiation on internal and external illuminance is excluded.

Daylight autonomy is the part of the year when a predetermined internal illuminance is exceeded during the working day. The relation between daylight factor illuminance and daylight autonomy is shown in Fig. 7, depicting the situation in Tallinn, Estonia, at latitude of 59.3°. On a cloudy day, the external illuminance is 15,000 lx. Illuminance of 300 lx or higher is ensured for around 40 % of working hours. With a daylight factor of 3, the illuminance of 300 lx would be ensured for about 65 % of working hours.

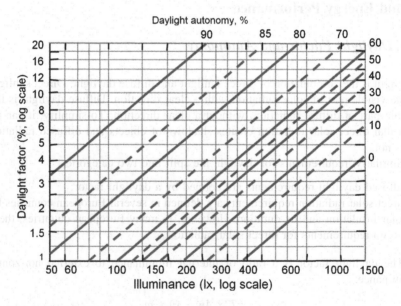

Fig. 7 Annual daylight autonomy in Tallinn 9.00–17.00

3.2 Thermal, Visible Light and Solar Transmittance of Window Glass

Window glazing has three focal features: thermal transmittance, visible light and solar transmittance. Thermal transmittance, or the U-value, W/(m^2 K) shows the rate of transfer of heat W/m^2 per one degree of temperature difference.

Since visible light is only a fraction of solar radiation, different characteristics to describe solar transmittance and visible light transmittance are used. The solar factor (g-value, also called total solar energy transmittance and solar heat gain coefficient) shows how much of the solar radiation falling on the window glazing enters the room, both directly through the glazing and through absorption into the panes, thus raising panes' temperature and transferring convective and radiative heat from the inner surface of the pane into the room. The smaller the g-value of glazing is, the less the solar radiation enters the room. For example, if the g-value is 0.4, then 40 % of solar radiation falling on the window will enter the room and 60 % will reflect, absorb and transfer away to outside.

The visible light transmittance of window glass τ_{vis} (–) describes the transmittance of visible light through the window, similar to the g-value. If the τ_{vis} is 0.8, then 80 % of the visible light falling on the window enters the room and 20 % will reflect or be absorbed. τ_{vis} applies to direct light, and to calculate the τ-value of diffused radiation, τ_{vis} is multiplied by 0.91.

Double-skin facades will reduce solar transmittance and visible light transmittance. For example, if the g-value of the external glass is 0.85 and the g-value of a window glazing unit is 0.5, then the total $g = 0.85 \times 0.5 = 0.43$. Total τ_{vis} is calculated accordingly. The maximum visible light transmittance of a double-skin facade can reach, with very clear glass, up to $\tau_{vis} = 0.87 \times 0.71 = 0.62$ (single-pane glazing unit + triple-pane glazing unit), which means that the $\tau = 0.51$. Without the external glass, that is, without a double-skin facade, the τ would be much higher, i.e., $\tau = 0.65$. To calculate the total thermal transmittance, all thermal resistances (the reciprocal of thermal transmittance) must be added together, but since external glazing is usually single pane and the air gap is ventilated (no resistance), in the approximate calculations, the thermal transmittance of the glazing unit can be used. This means that the double-skin facade does not affect the thermal transmittance of windows.

In terms of energy efficiency, glazing units with as low g-value as possible and as high a τ_{vis} as possible should be used. This enables the use of windows with a reasonable size, while at the same time ensuring good use of daylight. To some extent, g-value and τ_{vis} of windows are linked, and this is described by the selection of triple-pane glazing units, as shown in Fig. 8. There is also a certain interaction with thermal transmittance, since, in multi-pane glazing units, τ_{vis} is reduced (Fig. 9). Separately marked in the figures are so-called cold climate glazing units, thermal transmittance of which is 0.52–0.6 W/(m^2 K), and their τ_{vis} is between 0.59 and 0.71, with a g-value between 0.32 and 0.49.

Since a lower g-value ($g < 0.4$) also tends to decrease τ_{vis}, more efficient solar shading is achieved with external shading. For example, external blinds block ca 90 % of solar radiation ($g = 0.1$).

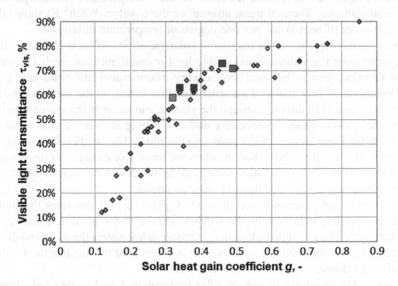

Fig. 8 Dependence of visible light transmittance on the g-value in triple-pane glazing units

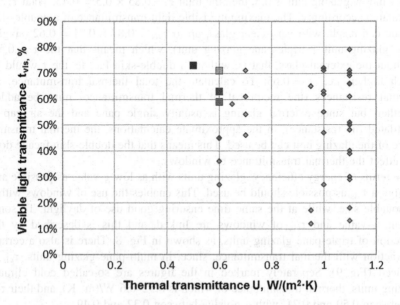

Fig. 9 Dependence of visible light transmittance on thermal transmittance in triple-pane glazing units

Glazing units with low thermal transmittance [$U < 1$ W/(m^2 K)] are characterized by condensation of vapour on the external surface of the glass. This usually occurs in the autumn, when the sky is clearing after rain. The air is very humid at that point, and the long-wave radiation from the outer pane surface to the sky cools the pane surface and this causes condensation or icing. Although it is merely a visual problem, it has become an annoyance. Window manufacturers offer several solutions to this problem. One proven solution is to use a hard selective coating (tin oxide SnO_2 or indium tin oxide (ITO), emissivity $\varepsilon = 0.15...0.2$) on the external surface of the external glass. External factors and window cleaning do not damage a hard selective coating, and it cuts 80 % of thermal radiance, which keeps the temperature on the external surface of the window higher, and thus, vapour condensation occurs more rarely.

Figure 10 shows required window-to-wall ratio as a function of visible light transmittance to ensure mean daylight factor of 2 in a typical office landscape. The daylight zone was estimated at 4 m, which would easily fit two consecutive work spaces. In the figure, the curve increases until the window covers all walls on top of the desk height (0.8 m). From then on, the daylight factor does not change, because the part of the window below the desk height does not significantly increase the amount of daylight, and it is not taken into account due to typical obstacles (furniture). Since the daylight factor is calculated in relation to diffuse sky radiance, it does not depend on the direction of the window.

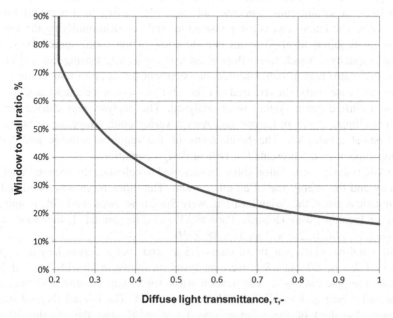

Fig. 10 Daylight factor dependence on window surface and the diffuse light transmittance factor. To calculate the diffuse light transmittance factor, the visible light transmittance factor τ_{vis} is multiplied by 0.91

In designing window sizes, a compromise between the view, daylight, low heat loss and cooling load should be found. Large window surfaces add heating and cooling loads and raise the need for an external solar shading. That regards the cooling load, internal solar shading does not work, as it allows solar radiation inside the room and even can increase cooling loads, as blocking storage to massive floor slabs.

4 Effect of Window Size on Cooling Capacity and Energy in Various Climates

The energy use of a building depends on the qualities of building envelope and the energy efficiency of the selected HVAC system. The properties of the windows are the most significant factor on cooling demand in modern offices, where energy-efficient light fittings and laptop computers are enabled [1, 2]. With good solar shading, the cooling requirement can be significantly reduced. The reduction in cooling loads also expands the variety of HVAC systems, which can be used in buildings. Low-temperature heating and high-temperature cooling air–water systems can be more easily introduced in such buildings where efficient solar shading is introduced [3].

During the design phase, it is important to make a difference between sensible cooling and total cooling loads, when air–water systems are considered. In air–water room air-conditioning systems, only sensible cooling load is covered with room units. The latent load is compensated in AHU by dehumidifying the supply airflow to required level to avoid condensation in the room space. Thus, the cooling capacity is much lower than when using, e.g., condensing fan-coil units, where the major part of dehumidification occurs in the fan-coil unit in the room spaces. In a case study, the required sensible and total cooling capacity and energy use of a chilled beam system were analysed. The analysis was carried out in different climate zones in Europe and Asia covering cold, temperate, subtropical and tropical conditions. The breakdowns of the required sensible and latent cooling capacities are presented in typical design conditions.

IDA-ICE energy simulation software was used to calculate the required cooling capacity and the energy use of an office room. The office room was simulated in six climatic regions: the Asian locations were Singapore, Seoul and Tokyo, and the European locations were Helsinki, Paris and Rome. The simulated office room area was 10.8 m^2 $(4.0 \times 2.7 \times 3 \text{ m}, L \times W \times H)$.

The window width was in all cases 2.5 m, and four different heights of the window 1.2, 1.6, 2.0 and 2.8 m were analysed, as shown in Fig. 11. The window sizes are presented in Fig. 1. The window was a triple-pane window with the solar factor (solar heat gain coefficient SHGC) of $g = 0.4$. The overall thermal transmittance (U-value) of the window was 1.1 W/m^2 K, and the U-value of the external wall was 0.3 W/m^2 K in all cases. The exterior wall was a concrete wall (heavy), and interior walls were plaster board structures (light). No heat transfer

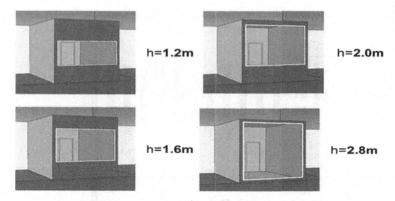

Fig. 11 Office building module division with longitudinal installation

between interior walls was assumed. Infiltration rate was 0.15 l/h during the cooling season.

Two occupants were considered to be in the simulated office room during the occupancy period from 9.00 a.m. to 6.00 p.m. The lighting load of 10 W/floor-m^2 and appliance load of 10 W/floor-m^2 were switched on during the occupancy. Fans operate from 7.00 a.m. to 8.00 p.m., providing a constant outdoor airflow rate of 2 l/s, floor-m^2. Room air temperature set point was 24 °C. The supply air temperature of the dedicated outdoor air system was 14 °C between April and August. During September to March, the supply air temperature was 19 °C (except in Singapore, the supply air temperature was continuously 14 °C).

In Asian cities, the required sensible cooling capacities of the office rooms were between 80 and 140 W/floor-m^2 (Fig. 12). In office rooms facing to the south and to the north, the sensible cooling capacities were at the level of 80 W/m^2. Only when fully glazed northward exterior wall was introduced (window height of 2.8 m), the required cooling capacity was at the level of 100 W/floor-m^2. In offices facing to the east and to the west, the maximum cooling capacity was 140 W/floor-m^2 with the fully glazed facade. When the window height was 2.0 and 1.6 m, the required cooling capacity was reduced to 110 and 100 W/floor-m^2.

In European locations' southward facades, the cooling demands were higher than in Asian (Fig. 13). In Northern latitudes, the vertical incident angle of the solar radiation is greater, and thus, the solar load is higher. The effect of the window height was more significant in European cities than in Asian metropolises. As a rule of thumb, it could be estimated that 40 cm higher full-width window increases the cooling capacity with 15 W/floor-m^2. The maximum cooling power in south rooms varied between 80 and 120 W/floor-m^2. By reducing the window height to 1.6 m, it is possible to maintain the set room air temperature using the cooling power of 80 W/floor-m^2. In the east and west facing office rooms, the maximum cooling capacity was 120 W/floor-m^2. When the window height was 1.6 and 1.2 m, the required cooling capacity reduced to 90 and 80 W/m^2. In the north-façade office room, the cooling capacity was 70–90 W/m^2.

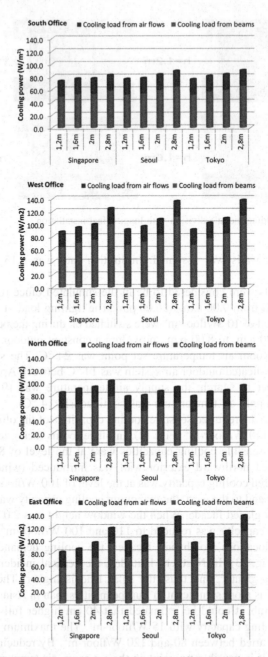

Fig. 12 Sensible cooling capacity in Asian cities using different window heights

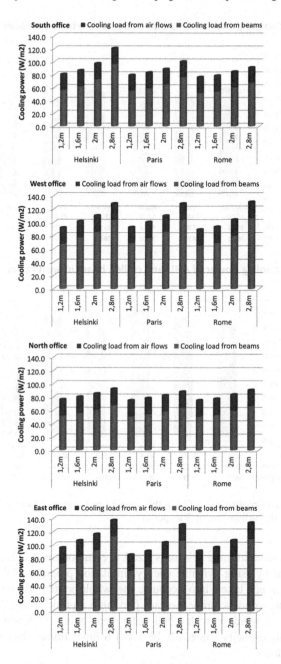

Fig. 13 Sensible cooling capacity in European cities using different window heights

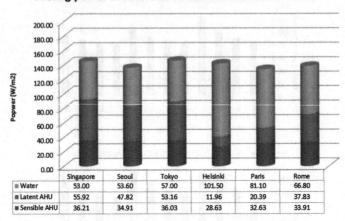

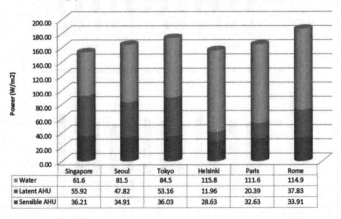

Fig. 14 Air and coil cooling capacities in the southern and western offices using a window of 1.6 m height and the airflow rate of 2 l/s per m^2

When an air-conditioning system is sized, it is important to calculate the actual cooling demand by using dynamic energy simulation program. If the effect of the thermal mass is not taken into account, the whole system is oversized. In the cooling demand, window properties are playing a significant role. If there is no solar shading or window with good solar heat gain coefficient (SHGC), the required cooling capacity can easily be 1.4–1.6 times higher than with the state-of-the art windows.

On the contrary of the local building codes in this study, the same U-values were used in simulations. However, it should be noted that heat transfer through structures is not so significant because of quite minor temperature difference between room space and ambient temperatures.

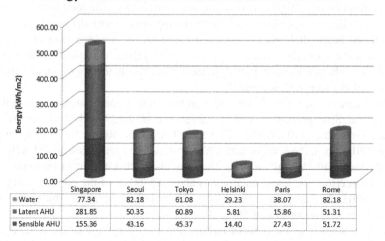

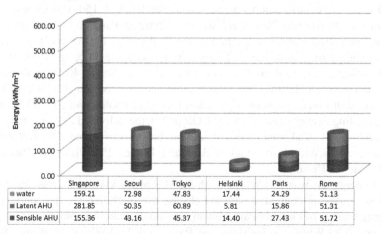

Fig. 15 Cooling energy use of air-handling units and water cooling of chilled beams

Introducing a window of 1.6 m height in south-façade office, the cooling capacities of the AHU and the chilled beam coil capacity were about the same 130–140 W/m^2, regardless of the location (Fig. 14). In west-façade offices, the cooling capacities were 160–180 W/m^2. The coil capacity was the most significant portion in the cooling capacities of the office rooms. In Europe, the sensible cooling of the chilled beam coil was 60–75 % of the office room. In Asia, the coil capacity was 40–50 %, respectively.

It is interesting to notice that the total cooling capacity of AHU and room system is about the same level in all analysed cities. In Asian and Southern

European cities where the solar load is lower than in the northern region, the latent load is, respectively, higher. It is important to notice that the total heat load is the basis for chiller sizing and room air-conditioning system with dry cooling principle like chilled beam shall always be sized based on the sensible load of the room space. On the contrary when sizing, e.g., split system or fan-coil room system, where the dehumidification partly or totally occurs in the room space, also the latent load is taken into account.

The location had a significant effect on the cooling energy demand in AHU. In the tropical Singapore, the cooling energy use of AHU with 2 l/s per m^2 was slightly more than 430 kWh/m^2. In Seoul, Tokyo and Rome, the cooling energy use of AHU was at the level of 100 kWh/m^2. Cooling energy demand of AHU was slightly above 40 kWh/m^2 in Paris and only about 20 kWh/m^2 in Helsinki.

In Asian cities, the latent energy was about 55–60 % of the cooling energy use in the AHU. In European cities, there was a larger variation: the ratio varied between 30 % (Helsinki) and 50 % (Rome).

With 2 l/s per floor-m^2 outdoor airflow rate, the energy use of the AHU and the chilled beam coils was about 510 kWh/m^2 in south-façade office room and about 590 kWh/m^2 in west-façade office room in Singapore (Fig. 15). In Seoul, Tokyo and Rome, the cooling energy need was at the level of 150–180 kWh/m^2. The cooling energy use was 70–80 and 40–50 kWh/m^2 in Paris and in Helsinki.

The latent load was about the half of the total energy use for cooling. The coil cooling energy use was of the total energy in south-façade office 15 % and in west-façade office 27 %. Coil cooling was about 45 % in Seoul, 40 % in Rome and 35 % in Tokyo. In Helsinki, the total cooling energy need was 40–50 kWh/m^2 in west and south-façade offices. The ratio of the coil cooling was 45–60 % of the whole cooling energy need. With the COP of 3.0, the delivered electric energy use of chiller (AHU and water coil of chilled beam) was about 180 kWh/m^2 in Singapore; 50 kWh/m^2 in Seoul, Tokyo and Rome; 25 kWh/m^2 in Paris; and 15 kWh/m^2 in Helsinki.

Fan energy use was 15 kWh/m^2 with the SFP of 1.7 kW/m^3/s. Thus, fan energy use was significant compared with the electric use of chiller. On the contrary, the pumping energy of chilled beam system was small. Specific pumping energy use was only 0.3 kWh/m^2 in Helsinki and Paris, and 2.0 kWh/m^2 in Rome, Seoul, Tokyo and Singapore.

5 Solar Shading and Examples of Facade Solutions

The purpose of solar shading is not only to block direct sunlight but also to ensure as much diffuse daylight as possible. During the spring and summer periods, the duration of direct solar radiation entering the room should be minimal, to avoid glare and to decrease cooling energy.

It is wrong to believe that large glass surfaces automatically mean well daylight rooms. If the issue of solar shading is not resolved, these buildings show that window curtains are closed around the clock, and spaces are illuminated by artificial light, as shown in Fig. 16.

Studies have shown people's "laziness" in handling curtains according to the need. Usually, spaces with no direct solar radiance have their curtains always open, and spaces with direct solar radiance have their curtains always closed. It has been noticed that office workers in the beginning adjust the curtains for the view and daylight, in accordance with the direct solar radiation effect, but later give up and leave the curtains constantly closed.

Closed curtains can be prevented by effective solar shading. Effective solar shading can be ensured by means of the following solutions:

- External blinds (lamellae);
- External overhangs;
- Double-skin facade with blinds or other shadings in between;
- Self-shading facade;
- Combinations of previous solutions.

Internal and external blinds are two distinct solutions. Blinds installed on the facade are an effective solution for blocking direct solar radiation. External blinds are quite widely used in Western Europe. Significant cooling load reduction is also achieved with blinds installed between window panes.

Internal rib curtains, roller blinds and other curtains could be called as emergency solution in terms of solar shading, because they let solar radiation inside and tend to increase rather than decrease room temperature.

External rib curtains or lamellae have wider ribs than usual (ca 5–8 cm). Their advantage is the possibility of automatic control depending on the amount of direct

Fig. 16 Solar shading of the facade is not resolved. Most of curtains are constantly closed. During daytime, spaces are lighted with artificial lighting, instead of daylight

solar radiation. With the lamellae system, it must be pointed out that when the solar altitude angle is ≤30, the lamellae are so far closed that seeing outside is impossible.

External overhangs blocking direct solar radiation are the most common and well-known type of passive architectural solution. Figure 17 illustrates main external shading solutions.

As a rule of thumb, a vertical overhang is more suitable for eastern and western facades, while a horizontal overhang is more suitable for southern facades. There is always the possibility to use a combination of horizontal and vertical overhangs. Necessary overhang dimensions depend on the location of the building—its geographical latitude. For example, at latitude of about 60°, the solar angle of incidence toward the vertical plane is the largest on June 21 at 12 p.m. and is about

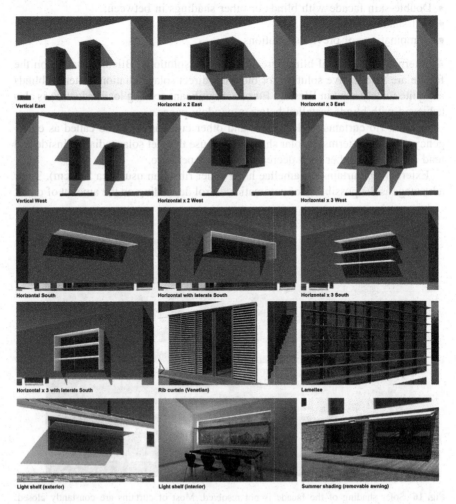

Fig. 17 Some examples of external overhangs and shadings

Table 2 External overhang options for blocking direct solar radiation on the southern side at 12 p.m. according to the length of the overhang in latitude 59°

12 p.m., noon	Solar angle of incidence (°)	Length of an overhang				
		40 cm	50 cm	60 cm	70 cm	100 cm
21 June	54.5	56	70	84	98	140
21 May/July	51	49	60	73	86	123
21 April/August	43	37	47	55	65	93
21 March/September	32	25	31	37	44	62
21 February/October	20	15	18	22	25	36
21 January/November	12	8	11	13	15	21
21 December	8	5	7	8	10	14

54°, as shown in Table 2. The same angle on December 21 is a mere 7°. If we wish to block direct solar radiation on the southern facade from the beginning of April until the beginning of September, then, within the specified period, the maximum solar angle of incidence is around 42° during the day. This means that the length of external shading would be longer than the width of the window (at 10 a.m. and 2 p.m., the solar angle is lower—39°). If a building has floor-to-ceiling-sized windows with a height of 3 m, it would mean that the overhang length should also be around 3 m.

Double-skin facades have been particularly popular in, for example, Germany, Austria, Finland, Sweden and Denmark. Between double-skin facades, rib curtains or lamellae are a very suitable solution, because they are quite well protected from the weather there.

Self-shading facades can be designed with inclinations, gradations and other architectural forms. For example, a southern facade can be built with a certain inclination. Possibility is an external overhang that reaches over the roof, which can be a solution for lower, up to two stories high, office buildings. The third and the most common option used in practice is a combination of the two aforementioned options. Such solution is shown in Fig. 18.

An interesting example of a self-shading facade is also a new office building in the Port of Tallinn, as shown in Fig. 19. When studied in a master's thesis, it was found that both southern and western glass facades of the building would be most effectively protected from direct solar radiation either by a overhang reaching 8 m over the edge of the roof, or by an option where the facade would be at a 45° angle towards the ground with a 4-m-long overhang to reach over the edge of the roof—to be possible to avoid direct solar radiation on the southern facade during working hours from March until September. Compared to the existing building, the cooling load would decrease by 14 % and annual cooling energy by 13 %. At the same time, the heating load would increase by 2 % and heating energy by 3 %.

The above-mentioned effective solar shading solutions can be combined in any suitable way. Figure 20 depicts an office building in Copenhagen, on the southern facade of which architectural elements were used, while the western part of the building has a double-skin facade, with rib curtains in between.

Fig. 18 A combination of self-shading façade with inclination and an external overhang

In conclusion, it can be said that, in terms of solar shading, the size and the shape of the window should be carefully through, since suggestions will be difficult and few in the case of large windows. There are not many options in solar shading a glass building. It should be shaded by means of vertical and horizontal overhangs. However, all curtains tend to remain closed in glass buildings, and in terms of heat and cooling, glass cannot compete with an opaque wall. The only realistic way for constructing a glass building is to use a double-skin facade. But if floor-to-ceiling-sized windows are designed, and adequate indoor climate is expected, this combination is not possible in the case of low and near-zero energy buildings.

Fig. 19 Tallink new office building in Tallinn

Fig. 20 Office building in Copenhagen, where architectural elements were used as a solution for the southern facade and a double-skin facade was used on the western side of the building

To ensure daylight and protection from overheating indoors, it is expedient to choose windows with as large a visible light transmittance as possible, with the size derived from the daylight factor. To prevent spaces from overheating, demand-controlled solar shading, operated automatically (+manual override), is the best solution. Decreasing the cooling load with window glass toning is ineffective. This requires larger windows to ensure daylight, which in turn again increases the cooling load. To exit this vicious circle, solar shading that can be controlled according to the need should be preferred.

6 Space Usage Flexibility in HVAC System Design

6.1 Changing Office Work

Recent studies have clearly proven the correlation between indoor air quality and the work performance of employees [4, 5]. Similarly, it has been demonstrated that the thermal conditions of the room have a significant impact on the productivity of work [6, 7]. Employee salaries and the potential change in productivity amount to many times the cost of a building technology system. The studies indicate that an investment in a better indoor environment is a profitable one, even with very minor productivity changes [8].

In addition to the indoor environment, the functionality of the workspace significantly affects the productivity of employees. Often a compromise must be made among the needs of the employee, team and organization when arranging workspaces. Addressing the interaction and privacy needs of employees, both of which are important considerations in organizations, is particularly challenging. In general, it can be stated that from the perspective of dispersing silent information (views, experiences, intuitions), fully autonomous workspaces do not support the business models of most companies. On the other hand, reducing the autonomy afforded by individual workspaces reduces acoustic privacy, which disturbs concentration.

Organizational changes in most companies are continuous and require flexible changes in work methods and workspaces. The traditional one-person office areas, or cells, and open offices, or hives, seen in traditional offices are today changing into spaces that are more suited to team work, referred to as dens or clubs (Table 3) [9]. In addition to this, information technology contributes to independence of time and

Table 3 Adapting space types and business processes in office buildings

Space	Interaction	Autonomy	Operation	Example
Hive	Low	Low	Customer service	Call centre
Cell	Low	High	Support tasks	Financial administration
Den	High	Low	Team work	Media
Club	High	High	Expert work	Consultancy

location, transforming offices more into meeting places for sharing information. The office space must be utilized efficiently, and therefore, a dedicated workstation is no longer deemed necessary for a worker who spends only a few hours a day at the office. Working at several workstations and at customer sites is becoming more common.

The supply airflow required in office buildings is 1.5–2 l/s per m^2, with the exception of meeting rooms and similar spaces. On the other hand, cooling (50–70 W/m^2) is required for attaining a good indoor environment. A beam system offers the possibility of achieving the above objectives within economical life-cycle cost limits. Therefore, the beam system has become one of the most common systems in office buildings. It should be borne in mind that the various systems should be applied according to the requirements in each space. For example, variable air systems (VAVs) should be selected for meeting rooms and displacement ventilation for auditoriums and common spaces. Thus, can an optimal overall solution be obtained with respect to expenses, energy use and environmental impact [10].

6.2 Flexibility in a Beam System

In design, ventilation beams are typically installed without a suspended ceiling in the room space, in the longitudinal direction. The basis for the system is the column spacing (e.g., 8.1 m), which can accommodate three one-person office rooms or two larger rooms (Fig. 21). This is possible because the beam is located at the side of the room instead of in the middle of the room module.

The ventilation ductwork should be designed for constant pressure, which enables demand-based airflow control in meeting rooms. Pressure adjustment is implemented per zone or by individual ventilation device. It should be taken into consideration in the management of airflows in a meeting room that the pressure

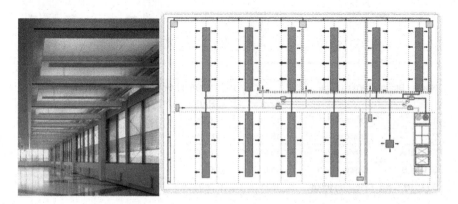

Fig. 21 Office building module division with longitudinal installation

level in a beam system is 60–100 Pa and the pressure level of a standard VAV is considerably higher than this. The variable air unit installed in the system must operate at the pressure level of the beams.

When the location of partition walls changes, the size of the room space and the supply airflow of the room device change as well. The typical office module division can result in three different cases: (1) one room device produces supply air for one room module, (2) it does so for 1.5 room modules, or (3) two room devices provide the required supply airflow for 1.5 room modules (Fig. 22). For maintaining a constant supply airflow (2 l/s per m^2), an adjustment is required.

Ready adaptability of airflow and space arrangements also increase the need to manage air distribution such that it reflects the various space solutions. It must be possible to reduce the total airflow from a beam in situations where, for example, the room device is close to a partition wall and the distance of the workstation from the wall is short (Fig. 22: room module 1). It should also be noted that owing to individual differences between people, some people perceive even low air velocities as a draught. This means an increased need to manage the individual room conditions.

It is important to remember that the majority (70–80 %) of the airflow distributed to the room is recycled indoor air induced by the ventilation beam through the thermal exchanger to obtain the required cooling effect. The supply airflow provided by the fans is only 20–30 % of the total airflow. This means that if the

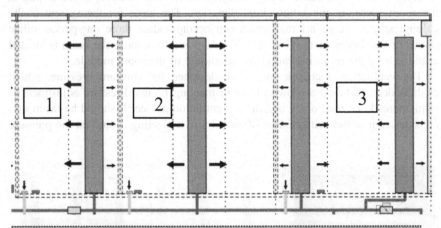

Room / Beam	Space Type	Primary airflow rate			
		Nozzles	HAQ	Total	
1 /	Office	15 l/s	5 l/s	20 l/s	2 l/s,m2
2 /	Office	15 l/s	15 l/s	30 l/s	2 l/s,m2
3 / Unit A	Office	15 l/s	0 l/s	15 l/s	2 l/s,m2
3/ Unit B	Office	15 l/s	0 l/s	15 l/s	2 l/s,m2
3 / Units A &B	Meeting room	15 l/s	0...45 l/s	15...60 l/s	6 l/s,m2

Fig. 22 Impact of room module division on the office and meeting room airflow

supply airflow is 2 l/s per m², the volume of air continuously recycled in the room is 8 l/s per m². Therefore, an efficient way of managing room space air velocities is to reduce the induction ratio of the room device to an appropriate level. Figure 23 presents the principle of operation for the induction ratio adjustment in a ventilation beam.

Figure 24 presents the impact of the induction ratio adjustment on the flow range of a CFD-simulated example room. The simulations indicate that induction ratio adjustment allows for reducing air velocities in the proximity of a window wall and at floor level. In Fig. 4, induction adjustment is used on both sides of the beam. In practice, only one-sided induction ratio adjustment can be used, which lowers the effect of induction ratio adjustment on the room temperature.

Case studies carried out in laboratories examined the significance of induction adjustment in a situation where beams installed in a suspended ceiling were perpendicular to the window. According to the measurements made, induction adjustment could significantly reduce air velocities in the proximity of the employees (Fig. 25) [11].

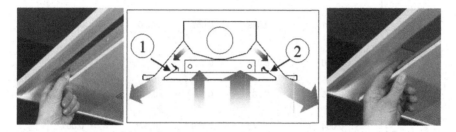

Fig. 23 Principle behind induction ratio adjustment *1* induction ratio adjustment on and *2* induction ratio adjustment off

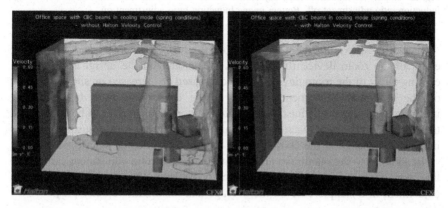

Fig. 24 Impact of two-sided beam induction ratio adjustment on the air velocity for a sample room (threshold velocity: 0.25 m/s) and its temperature. *Left* Induction ratio adjustment off. *Right* Induction ratio adjustment on

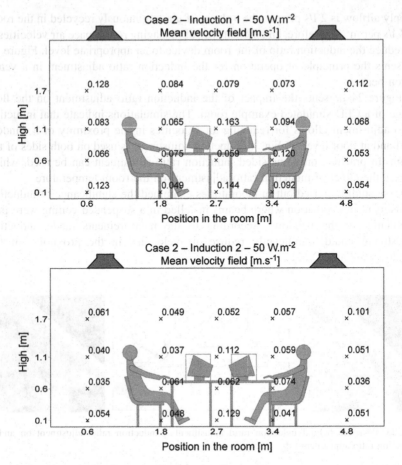

Fig. 25 Impact of induction ratio adjustment on air velocity in laboratory tests. *Top image* Induction ratio adjustment off. *Bottom image* Induction ratio adjustment on

Induction ratio adjustment simultaneously affects air velocity and the cooling effect of the room device (by 5–15 %). Depending on the design solution, the cooling effect of the room device can be decreased or increased on site, depending on the induction adjustment position in the design situation. In comparing induction adjustment to adjusting the water flow in the beam, it can be determined that when the goal is to reach the same air velocity, water flow adjustment results in a greater change in the cooling effect than induction adjustment does [12].

In addition to the maximum air velocity, induction ratio adjustment can reduce the average room air velocity. Figure 26 presents the impact of induction adjustment in a sample case on the average and maximum velocity [13].

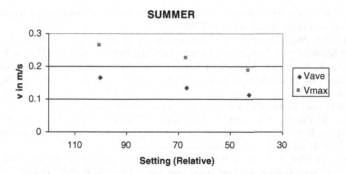

Fig. 26 The impact of the total airflow supplied in the room on the maximum and average air velocity in the occupied zone as a function of the relative change in the induction ratio adjustment

6.3 Discussion of Flexibility

In a modern office environment, balance is sought between work performed by individuals and in interaction between employees. It must be possible to appropriately combine various work methods, so partition walls and workstations should be flexibly adaptable if they are to best meet the business needs of individual customers.

Adaptability of office space is one of the central requirements in the design of a beam system. The systems must be adjustable to address changed loads and partition wall positions. In design of a room system, this flexibility means taking account of supply airflow, cooling effect and supply air device throw length pattern changes. In addition, adjustability entails requirements concerning room automation, actuators and sensors. It must be possible to adjust the airflow supplied by the room device, which enables managing the maximum and average velocity in the occupied zone and thus reducing the draught risk.

References

1. Virta M, Itkonen H, Mustakallio P, Kosonen R (2010) Energy efficient HVAC-systems and building design. In: Proceedings of Clima 2010, 10th REHVA world congress. sustainable energy use in buildings, 9–12 May Antalya, Turkey
2. Wilkins C, Hosni M (2011) Plug load design factor. ASHRAE J 53:30
3. Virta N, Hovorka F, Litiua L, Kurniski J (2012) HVAC in sustainable office buildings. REHVA Design Guide Nro 16
4. Kosonen R, Tan F (2004) The effect of perceived indoor air quality on productivity loss. Energy Buildings 36:981–986
5. Wargocki P, Wyon DP, Baik YK, Clausen G, Fanger PO (1999) Perceived air quality, SBS symptoms and productivity in an office at two pollution loads. In: The 8th international conference on indoor air quality and climate, Edinburgh

6. Kosonen R, Tan F (2004) Assessment of productivity loss in air-conditioning buildings using PMV index. Energy Buildings 36:987–993
7. Wyon DP (1996) Individual microclimate control: required range, probable benefits and current feasibility. In: Proceedings of indoor air '96, Institute of Public Health, Tokyo
8. Hagström K, Kosonen R, Heinonen J, Laine T (2000) Economic value of high quality indoor air climate. In: Proceedings of healthy building 2000 conference, 6–10 Aug 2000, Espoo, Finland
9. Flexibility Ltd., Cambridge, UK, at http://www.flexibility.co.uk/
10. Kosonen R, Hagström K, Laine T, Martiskainen V (2003) A life-cycle study of an office building in Scandinavian conditions: a case-study approach. In: Proceedings of healthy building 2003 conference, 7–11 Dec 2003, Singapore
11. Zhoril V, Melikov A, Kosonen R (2006) Air flow distribution in rooms with chilled beams. In: 17th air-conditioning and ventilation conference 17–19 May 2006, Prague
12. Ruponen M, Streblow R, Mustakallio P (2005) Room velocity control for a room ventilation device. Paper presented at 8th REHVA world congress CLIMA 2005, 9–12 Oct 2005, Lausanne, Switzerland
13. Streblow R (2003) The effect of secondary volume flow in induction units on room air velocities. Hermann-Rietschel-Institut fur Heizungs-und Klimatechnik. Technical University of Berlin, Diploma Thesis no. 343, 2003

nZEB Case Studies

**Jarek Kurnitski, Matthias Achermann, Jonas Gräslund,
Oscar Hernandez and Wim Zeiler**

Abstract There already exist pilot projects across Europe, which may be called nZEB buildings. They are not easy to compare because of variation in the definitions and performance levels—nZEB definitions have not yet been available when these buildings have been designed. It is important to check which energy uses are included in the calculated and measured energy performance and are the results reported as delivered or primary energy. Tenant's electricity (appliances, lighting) is often not included. Simulated primary energy is typically between 50 and 100 kWh/(m^2 a) for high-performance nZEB buildings with on-site renewable energy production if all energy uses are included. Such buildings may be extremely complicated, e.g. control of mixed mode ventilation or integrated energy supply solutions with storage and many operation modes. "Overkill" complexity may have implications on operation and maintenance as nobody cannot manage systems, which are not easy to control and operate. Another trend that can be seen is simple and reliable solutions based on high-performance components and careful system design and fitting with building properties. In the following, five nZEB buildings across the Europe are reported. Three of them are with measured energy data, indicating that strict targets are not always easy to achieve; however, some deviations have been caused also by unrealistic energy calculation input data. Technical solutions used show that there are many alternative ways to achieve high performance; however, strong differences in design bases and basic solutions in similar climate may also show that optimal solutions are always not known or found. Four first case studies are mostly focused on description of technical solutions, but the last one also describes the tuning of systems needed in the first year of operation to achieve the targets.

J. Kurnitski (✉) · M. Achermann · J. Gräslund · O. Hernandez · W. Zeiler
Tallinn University of Technology, Ehitajate tee 5, 19086, Tallinn, Estonia
e-mail: jarek.kurnitski@ttu.ee

J. Kurnitski (ed.), *Cost Optimal and Nearly Zero-Energy Buildings (nZEB)*,
Green Energy and Technology, DOI: 10.1007/978-1-4471-5610-9_8,
© Springer-Verlag London 2013

1 Elithis Tower in Dijon, France

Elithis Tower, located in Dijon, France, provides strong evidence that net zero-energy office buildings are achievable in near future. The building, which was designed by Arte Charpentier Architects, also produces six times fewer greenhouse gas emissions than traditional office structures.

The Elithis Tower is an experimental and demonstration building. A lot of R&D are being done in order to improve energy performance. The principal objective was to erect a nZEB building with architecture fitted to an urban environment.

An environmental protocol was signed by all the permanent co-owners of the Elithis Tower in order to ensure to lowest impact between user's behaviour and the building. The energy production of the building in kilo Watt per hour and the greenhouse gas compensation is permanently projected to the advertising board on the public road.

Thermal comfort, indoor air quality, and energy use are being constantly monitored with 1,600 data points installed all over the building. In addition, occupant surveys are done for the users. Users are asked to fill in a questionnaire at the same time as the environmental variables are being recorded through the BEM system. The study began in June 2010 and the first results report a general comfort level of 72 % (winter season), including thermal and visual comfort and indoor air quality (Fig. 1, Table 1).

Fig. 1 Elithis Tower in Dijon, France

Table 1 General data of Elithis Tower

Elithis Tower	
Financing	ADEME, Conseil Regional de Bourgogne
Net construction costs	EUR 7 millions, 1,400 €/m² (equals the cost for a standard building in France)
Project team	Elithis Ingénierie, ARTE Charpentier
Building type	Office
Net floor area	4,500 m²
Gross floor area	5,000 m²
Gross volume	167,500 m³
Mean occupant density	15 m²/person (overall average)
Occupied hours	2,450 h
Climate data	
Design outdoor temperature for heating	−11 °C
Design outdoor temperature and RH for cooling	32 °C/38 %
Heating degree days (base temperature)	2,650 degree days (base 18 °C)

1.1 Building Description

The main aim of the building is to use passive means and natural resources such as sun and wood to achieve thermal and visual comfort in the building. In order to improve the best performances in natural lighting, the Elithis Tower was designed in an open-plan distribution. Unfortunately, this configuration was not adopted all over the building (medical services). Most part of the offices are in an open-plan distribution. But for the other offices, the glass wall and insulated wall division were installed. The open-plan distribution could ensure the best internal air distribution, and this solution gives the possibility to perform the air contact with the walls and to reduce the energy requirements for the cooling and heating (Fig. 2).

The Elithis Tower is composed of nine levels and one technical level (HVAC system). The height is 33.5 m. Four levels are occupied by Elithis engineering, and the others by the Ademe (Departmental Agency of Energy Management), radiological services, a restaurant, and other civil engineer companies.

The building has a central core made of concrete and the facades are made of wood and recyclable insulation (cellulose wadding). The surface fenestration is about 75 % of the facades. The windows are double-glazed with an argon air space. The thermal mass of the building can be considered as medium because the central core only is the exposed concrete.

1.2 Compactness and Solar Shading

Elithis Tower has very compact rounded shape effectively reducing building envelope area. The architecture was carefully studied in the design. The building envelope area of the Elithis Tower is about 10 % less than in a conventional tower.

Fig. 2 Indoor view of Elithis Tower

Reducing the surface has a positive effect regarding heat losses and solar gains. Similarly, the exposure to the wind is reduced so the infiltration can be better controlled. In the same time, the air distribution in the mixed ventilation mode can be more homogeneous thanks to the rounded shape (Table 2).

In order to combine natural light, avoid glare and reduce solar gains, a special solar shading shield was designed by the Elithis engineers and architects, Fig. 3. This passive system gives to the building the necessary natural light and the solar glare protection in summer and mid-season, while excess heat is utilized to heat the building in winter. The system was carefully studied in order to retrieve the necessary solar energy during the winter season and to protect the building during the hot periods.

Table 2 Building envelope data

Building components	
Window U-value	1.1 W/(m^2 K)
Window g-value	0.4
Exterior wall U-value	0.32 W/(m^2 K)
Base floor U-value	0.39 W/(m^2 K)
Roof U-value	0.22 W/(m^2 K)
Structural frame	Heavy weight (concrete and steel)

Fig. 3 Solar shading of the facade

1.3 Ventilation Strategy

The building is ventilated by mechanical supply and exhaust system with heat recovery controlled by the BEM system in order to comply with the French ventilations standards codes (25 m³/h per person in offices). The ventilation system is operated in three modes depending on the season.

For typical heating season operation (outdoor temperature higher than 0 °C), operation with controlled heat recovery is used to heat up supply air with heat recovered from extract air, Fig. 4. Heat recovery is controlled/bypassed so that

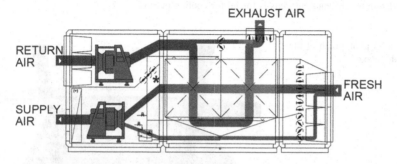

Fig. 4 Ventilation operation with controlled heat recovery during typical outdoor temperatures in the heating season

supply air temperature is between of 16 and 18 °C. The full heat recovery operation is used for extremely cold or warm outside conditions (less than of 0 °C in
winter or higher than 26 °C in summer).

In the mid-seasons (spring and fall) and summer operation, the triple flow mixed
mode system (Fig. 5), which is an Elithis innovation, is used. It gives the possibility
for ventilative cooling with fresh air intakes and central atrium exhaust ventilation
in order to cool the building. Thirty-two air valves in facades per level are used to
have additional intake air, Fig. 6. In this mode, air handling units are operated
together with intake air from facades and low-pressure atrium exhaust fans.

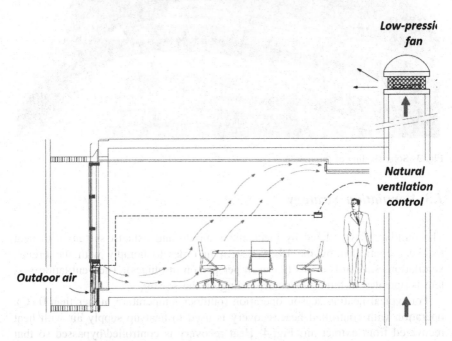

Fig. 5 In the mixed mode operation, façade intakes and low-pressure atrium exhaust fans are
used. This is used for the night-time ventilative cooling and, in the midseason, when the
ventilation by air handling units is forced for cooling purposes

Fig. 6 A photograph of the
façade intake

The third operation mode is the free cooling. Air handling units are stopped, and atrium exhaust is used in order to ventilate the building in night summertime. In this mode, the building can be ventilated with low-pressure central atrium exhaust ventilation. The 32 air valves are opened in order to ventilate the building with two or three times higher flow rate than the design airflow rate.

1.4 Lighting System

In natural lighting, increased rate of the glass surface reduces energy use needed for artificial lighting. The passive solar shading of the Elithis Tower protects the users from the direct solar radiation and provides an excellent natural lighting for the office tasks avoiding the glare problems.

Light fittings in the ceiling provide the average lighting (300 lux-French building standard codes) over the entire office space. For the low lighting outdoors levels, at night or very cloudy days, motion sensors were coupled with lighting sensors. This solution provides the perfect compromise between energy use and lighting requirements. Installed lighting power is only about 2 W/m^2 of electrical energy. For tasks requiring a higher level of illumination, task lighting with "nomadic lamps" is used. All this is controlled by the BEM system.

1.5 Heating and Cooling System

The major part of the heating needs is covered by solar and internal heat gains. For the rest of heating needs, one very low-power wood boiler provides the heat in order to ensure the thermal comfort. A second one wood boiler is used only for backup. This system is used to maintain the 21 °C room temperature all over the building.

The triple flow ventilation system covers the most important part of the cooling needs. When room temperatures reach 26 °C, a cooling system consisting of adiabatic unit and heat pump are started to operate; Fig. 7. This heat pump system with a high EER of (EER = 11) provides air-conditioning of the building. It is in two stages. The first one is an adiabatic process; the heat is evacuated by the water evaporation. The second stage of the heat pump is only needed to operate for extremely outside weather conditions (outdoor temperature higher than 30 °C).

Chilled beams of a rectangular cassette size are use as room conditioning units both for cooling and heating and ventilation supply air, Fig. 8. Chilled beams are induction devices circulating room air through the coil. Circulating airflow is induced by supply air nozzle jets integrated into chilled beams. Thirty-two chilled beams are installed per level and controlled by the BEM system.

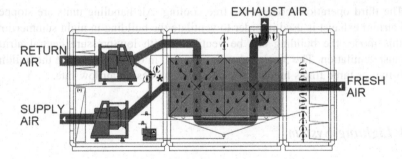

Fig. 7 Ventilation with adiabatic cooling in summer operation

Fig. 8 Chilled beams cassettes and lighting installation

1.6 Water Management

Water management is written in the policies of the Elithis Tower. A rain water recovery system is used to supply the toilets of the building. All fixtures and fittings such as sink faucets and toilets aim to very low water consumption in order to preserve the water resource.

1.7 Energy Performance

The energy concept of Elithis Tower is to balance the primary energy of all energy uses with the PV electricity generation and the user behaviour. A building by itself cannot be nZEB without a good operation and maintenance and users behaviour. The Elithis Tower has a very low ratio of installed PV area to the floor area. The very low energy use of the building is balanced by only 500 m^2 of photovoltaic's roof panels. The PV panels are installed with a horizontal inclination in order to maximize the generation.

An energy management system with 1,600 data points allows the control and the management of all technical systems (HVAC, lighting, elevators). Many energy metres are installed in all the building, to make it possible to know energy use on the system and component level. Simulated and measured energy performance of the building after the first year of operation is shown in Table 3.

The highest component in the energy balance are the appliances (plug loads), which include all user electricity, i.e. computers and other office equipments, cafeteria and also data servers. This component shows also highest deviation from the design value when all other components follow well-designed values. The differences between the theoretical patterns and the reality can explain this difference. As the user behaviour has been the most important reason to explain the differences in the energy balance of the building, Elithis Engineering is currently analysing the problem and there are many changes planned to be implemented in order to reduce that energy use.

Measured total primary energy use for the first year of operation year has been 63 kWh/(m^2 a) per net floor area, 57 kWh/(m^2 a) per gross floor area as calculated according to French standard, which is 33 kWh/(m^2 a) higher than designed, due to higher energy use of appliances.

Table 3 Simulated and measured energy performance of the building after the first year of operation. All specific values are per *gross floor area*

	Delivered and exported energy, kWh/ (m^2 a)	Design phase Primary energy factor	Primary energy, kWh/ (m^2 a)	Measured 2009 Primary energy, kWh/ (m^2 a)
Space, water and supply air heating, wood boiler	3.3	0.6	2.0	6.3
Cooling, electricity to heat pumps	4.1	2.58	10.6	6.2
Fans (HVAC)	5.1	2.58	13.1	14.1
Pumps (HVAC)	0.4	2.58	1.1	2.6
Lighting	4.1	2.58	10.5	9.5
Elevators	1.4	2.58	3.6	3.6
Appliances (plug loads)	9.4	2.58	24.2	54.6
PV power generation	−16.0	2.58	−41.3	−40.2
Total	**12**		**24**	**57**

The primary energy values reported include all energy use in the building, such as cafeteria, data servers, and all other activities in the building. Even the monitored primary energy value of 57 kWh/(m^2 a) is higher than designed, it places the Elithis Tower very close to high-performance net zero-energy building. The design value, not reached during the first year of operation, will remain the main objective in future operation.

1.8 Experience from the Operation

After nearly 2 years of operation, some improvements have been made or forecasted:

- At the beginning, the electricity used to light the stairways was higher than the electricity for the elevators. The problem was in the stairways lighting control, which proved to be very important because there is no natural lighting. Today, a new lighting programming is studied to solve the problem.
- The energy use predicted for the appliances was underestimated. The lesson is learnt, and in future, this will need more careful prediction. At the beginning, the device cut-off computer power was not used as expected and an awareness protocol was implemented in order to reduce the electricity use. Today, the systems seem to work and an energy reduction has been achieved.
- Occupants and visitors of the Elithis Tower are satisfied. The general feeling is very satisfactory because the environment is very attractive compared with other buildings.

2 Ympäristötalo in Helsinki, Finland

The Environment Centre building Ympäristötalo of City of Helsinki is a good example of the exemplary role of the public sector. It shows the best energy performance of an office building ever built in Finland. Total primary energy use of 85 kWh/(m^2 a) including small power loads is expected to comply with future nearly zero-energy building requirements. The building is also highly cost efficient, and nZEB-related extra construction cost was only of 3–4 % (Fig. 9, Table 4).

2.1 Energy Performance

The building has a high-quality building envelope, south facades being double facades with integrated PV cells providing effective solar protection at the same time. All the buildings, except the atrium space, are air-conditioned with effective

Fig. 9 Ympäristötalo in Helsinki, Finland. South facades are double skin facades with integrated PV. All façade are different because of daylight and solar shading considerations. *Photograph* Mari Thorin, Rhinoceros Oy

Table 4 General data of Ympäristötalo

Ympäristötalo, construction year 2011	
Construction management	City of Helsinki, PWD-construction management (HKR-Rakennuttaja)
Owner	City of Helsinki, environment centre
Construction costs	16.5 million € (2,430 €/m²)
Estimated nZEB extra construction cost	0.5–0.7 million € (70–100 €/m², 3–4 %)
Heated net floor area	6,390 m²
Gross floor area	6,791 m²
Occupants/mean occupant density	240/25 m²/person (overall average)
Architect	Ab Case Consult Ltd., Kimmo Kuismanen
HVAC design	ClimaConsult Finland

integrated balanced ventilation and free cooling system with passive and active chilled beams. All the cooling is from boreholes, which water is directly circulated in air handling units and chilled beams. Heating systems is based on district heating and water radiators. Highly significant energy efficiency measures are large air handling units and ductworks enabling low-specific fan power, combined with demand-controlled ventilation in most of rooms except cellular offices, and effectively controlled lighting. The simulated energy performance is shown in Table 5. On-site renewable energy production of 7.1 kWh/(m² a) PV power

Table 5 Simulated energy performance (all values per net floor area)

	Energy need, kWh/(m² a)	Delivered and exported energy, kWh/(m² a)	Energy carrier factor	Primary energy, kWh/(m² a)
Space and supply air heating	26.6	32.2	0.7	22.6
Hot water heating	4.7	6.1	0.7	4.3
Cooling	10.6	0.3	1.7	0.5
Fans and pumps	9.4	9.4	1.7	16.0
Lighting	12.5	12.5	1.7	21.3
Appliances (plug loads)	19.3	19.3	1.7	32.7
PV		−7.1	1.7	−12.0
Total	**83**	**73**		**85**

generation and 10.6 kWh/(m² a) free cooling from boreholes has significant effect on achieved total primary energy value of 85 kWh/(m² a). Typical to nZEB buildings, the highest primary energy component is the small power loads.

2.2 Compactness and Solar Shading

The building has a reasonable compact massing, and excessive glazed areas are avoided. The main facade to south is accomplished as a double facade in order to provide effective solar shading and to serve as mounting for PV panels. Architectural key solution for compactness is the grouping of office rooms around two inner courtyards, Fig. 10. These courtyards have insulated roofs with some vertical windows (can be shown from Fig. 11) and received daylight through glass façade.

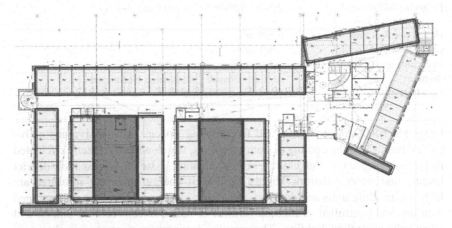

Fig. 10 Grouping office rooms (shown with *yellow colour*) around closed inner courtyards (*with red*) increases the depth of the building and results in improved compactness

Fig. 11 Atrium space of the building has no installed cooling and overheating is avoided by opening *bottom* and *top* windows. Motorized openable widows are to be open manually when needed and will be closed by weather station control (wind, rain, and temperature control) automatically

Window area is only 23 % of the external wall area, but the double main facade still provides an outlook of a glass building from major direction. The double facade is open from bottom and has motorized ventilation openings on the top. In total, there are about 30 motorized openings/windows used for the double facade and atrium excess heat removal and some of them are also used for smoke removal. The openings are to be open when needed (manual control from reception), and they will be closed by weather station control, based on wind, rain and temperature, automatically. Windows have blinds between panes.

2.3 Energy Supply

The building is connected to Helsinki district heating system. District heating is used for hot water and space heating through central air handling units and hot water radiators.

All cooling need is covered with free cooling from borehole water. The borehole system consists of 25 boreholes each 250 m deep. A simple borehole cooling system with a circulation pump and a water tank serves both the central air handling units and chilled beam units installed in offices and other spaces. Boreholes are sized to provide 15 °C supply design temperature (return 20 °C) to the water tank at dimensioning conditions (normally, the borehole water temperature is lower). Air handling units' cooling coils and chilled beams network are sized to 16/20 °C design flow temperatures from this water tank.

South facade of the building has a double facade with vertical PV panels, and some panels are also installed on the roof, Fig. 12. The total installed PV power is 60 kW (570 m^2) that provides about 17 % of electricity use of the building.

2.4 Ventilation and Air-Conditioning System

The building has an air-conditioning system with mechanical supply, exhaust ventilation and chilled beams, Fig. 13. There are three main air handling units (Fig. 14) and four risers with zone dampers for each floor. Separated exhaust fans for toilets are not used and are replaced with a small 0.5 m^3/s air handling unit with rotary heat exchanger. The main large air handling units of 2.4, 4.2 and 4.0 m^3/s have heat recovery temperature ratios of 80, 79 and 78 %, respectively. The rest, smaller air handling units have temperature ratios of 80–81 %. Ventilation system is balanced so that design supply airflows equal to design extract airflows.

Outdoor air is filtered and heated or cooled in central air handling units and supplied to rooms. Supply air is heated in the central air handling units partly with heat recovered from extract air and partly with heating coils. When cooling is needed, supply air is first cooled in the central air handling units and then cooled further in the chilled beam units.

2.5 Room Conditioning Solutions

All open-plan and cellular office spaces have room conditioning with active or passive chilled beam units installed in the ceiling and controlled by room temperature controllers, Fig. 15. Air volume flow rate is kept constant (constant pressure CAV). Rooms are heated with hot water radiators controlled by thermostatic radiator valves.

Fig. 12 PV installation on the double facade to south serving also as an effective solar shading

Active chilled beams are used in cellular offices and passive chilled beams in other rooms. Passive chilled beams allow to use cooling also during nights and weekends when ventilation is switched off. This reduces peak cooling loads to 40 W/m^2 that is important in the free cooling system with limited capacity.

Ventilation in the meeting rooms, lobby and workshop areas is controlled by CO_2 and temperature sensors. VAV dampers are used and airflow rates in the meeting rooms are controlled between 0 and 4 l/(s m^2). Office rooms have Constant Air Volume (CAV) ventilation of 1.5 l/(s m^2). The major part of cooling and heating are supplied by the water systems (beams and radiators, respectively).

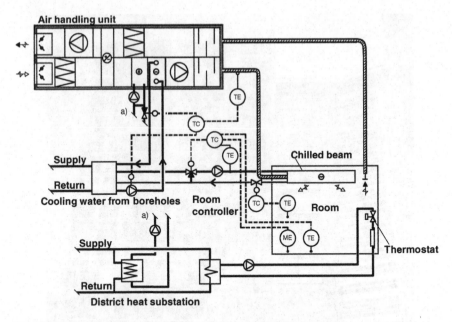

Fig. 13 Air-conditioning system with mechanical supply, exhaust ventilation and chilled beams

Fig. 14 Large air handling units and ductworks have been used to achieve as low-specific fan power as 1.4–1.6 kW/(m³/s) for offices and similar spaces and 1.8 kW/(m³/s) for VAV air handling unit serving meeting rooms

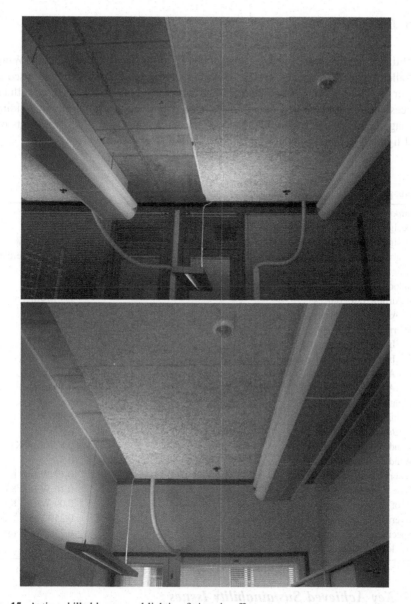

Fig. 15 Active chilled beams and lighting fittings in offices

Supply air temperature is extract air temperature compensated and is set between 17 and 22 °C. The supply air temperature is controlled by adjusting the rotation speed of the regenerative heat exchanger and the water flow control valves of the heating and cooling coils.

2.6 Lighting System

Lighting systems uses lighting fittings of T5 fluorescent lamps with 7 W/m^2 installed power. Occupancy sensors and photocell-controlled dimming are used in larger rooms, and occupancy sensors and manual dimming are used in cellular offices. Lights and chilled beam units have a communication link to building management system (BMS). Outside normal office hours, the BMS sets lights off and lighting demand is controlled by IR motion sensors.

Table 6 Technical data of Ympäristötalo

Outdoor climate data	
– Design outdoor temperature for heating	−26 °C
– Design outdoor temperature and RH for cooling	28 °C / 50 %
– Heating degree days (base temperature 17 °C)	3,952 degree days
Indoor environmental quality targets	
– Indoor air quality	
– Airflow rate, offices	1.5 l/(s/m^2)
– Airflow rate, meeting rooms	4 l/(s/m^2)
– Thermal environment	
– Indoor temperature, heating season	21 °C
– Indoor temperature, cooling season	25 °C
– Air velocity, winter	0.14 m/s
– Air velocity, summer	0.20 m/s
– Lighting	
Illuminance level	300/500 lx
– *Building envelope*	
– Window U-value	0.8 W/(m^2 K)
– Window g-value	0.3
– Exterior wall U-value	0.17 W/(m^2 K)
– Base floor U-value	0.16 W/(m^2 K)
– Roof U-value	0.09 W/(m^2 K)
– Average U-value of the building envelope	0.259 W/(m^2 K)
– Specific heat loss per net floor area H/A	0.276 W/(K m^2)
– Air leakage rate at 50 Pa	0.56 ach

2.7 Key Achieved Sustainability Issues

The best energy performance of an office building ever built in Finland. Total primary energy use of 85 kWh/(m^2 a) including small power loads is an half of a code requirement of 170 kWh/(m^2 a) and is expected to comply with future nearly zero-energy building requirement.

Very good indoor climate quality according to Finnish Classification of Indoor Environment.

Well-controlled construction cost of 2,430 €/m^2 is roughly a cost for standard office building in Finland, including only 3–4 % extra energy performance-related cost (Table 6).

3 IUCN Headquarter in Gland, Switzerland

The IUCN, as an international organization which is active all over the globe to preserve the natural environment, has set a high target for his extension of the Swiss headquarter in Gland.

Based on the wish of IUCN to create a showcase of sustainable construction and high efficient building technology, the interdisciplinary team went to work in 2006. The building finally was inaugurated in the spring of 2010. It complies with the Minergie-P-ECO and is aspiring the American LEED Platinum label. The key factor of success for the realization was the interdisciplinary collaboration. The close collaboration between architects and specialized engineers has made it possible to conciliate aesthetics, energetic performance and high flexibility for occupants with a very tight budget (Fig. 16).

Fig. 16 North-east façade of IUCN headquarter building

3.1 Interdisciplinary Design: A Key Factor for an Efficient Building

The starting point for a successful energy-efficient structure is an architectural concept, which takes into accounts passive solar heat gains and thermal losses. An optimized primary energy balance has been sorted through an iterative process changing the thermal performance of the envelope as well as the fraction of glazing and opaque wall parts and their thermal performances. The result of this optimization can now be identified with the work done: a relatively low rate of glazing compared to the surface of facades, a wall thickness of 35 cm, a high performing triple glazing as well as outside corridors for sun protection in summer and as emergency exits for users in case of fire (Fig. 17).

A key element of this optimization was the glazing, which strongly influences the cooling needs and the comfort of users. The 25 % glazing ratio of the facade can limit power peaks. To improve management of natural lighting without risking overheating due to solar radiation, movable blinds that are rolled from bottom to top direction, were used.

Fig. 17 Inner courtyard of the building

3.2 Energy from the Basement and Sun

Thanks to the thermal performance of the envelope, the heating need is very low. There is still necessary to heat the supply air of ventilation and domestic hot water. Requirements for space heating are secondary. Mainly because of an administration-bent working, the cooling need is by contrast predominant. It was then necessary to use a renewable source for cooling energy. Geothermal energy provided the answer. With a field of geothermal wells of a depth of 150 m, 30 % of cooling needs can be met by passive cooling. Cooling energy is produced by the reversible heat pump only when the free cooling reservoir is exhausted. Through the dissipation of heat in the ground, in the second part of the summer, heat warms the ground in order to optimize the performance of the heat pump in the following winter. In parallel with the heat pump connected to the geothermal probes, a heat pump on the exhaust air was installed to preheat the air of the decentralized air intakes. This heat pump is also reversible and can cover smaller cooling needs of the fan coils without disrupting the geothermal free cooling storage.

A 1,400 m^2 photovoltaic installation on roof covers the annual electricity needs. Seasonal overproduction is fed back into the electrical grid (Fig. 18).

Fig. 18 PV installation on the roof

3.3 Demand-Controlled Ventilation

Given that the occupation of areas of work is very varied, constant-flow ventilation would consume too much electrical energy and a traditional VAV facility would be too expensive. The adopted solution consists of small floor mounted decentralized outdoor air units, contributing independently to the ventilation and thermal comfort of users. For the entire administrative area, except for large conference rooms, decentralized units have been positioned close to the facades at floor level. These units (marked as AIRBOX in Fig. 2) are equipped with an air intakes from facade, a filter unit, a fan and a heating/cooling coil. The units operate only with outdoor air, and there is no air circulation. They are controlled according to CO_2 in the room air. The CO_2 sensor is located at the exhaust damper, integrated into a multifunctional panel mounted on the ceiling. Each ventilation unit is connected to an exhaust damper, both attributed to one facade frame. This system avoids a complete supply air ductwork. It allows a much easier routing for technical installations. On the other hand, an air quality management based on demand is possible. If the CO_2 is high, the ventilation starts, and if the air is clean again, the fan stops. If users are not present, two air flushes per day allow to keep a minimum fresh and good air quality.

The multifunctional ceiling panel, serving at the same time as thermal activation of the ceiling, as acoustic element and as light fixture, also includes an extract air terminal, Fig. 19.

Fig. 19 Ceiling panel for heating and cooling, with integrated extract air terminal and lighting

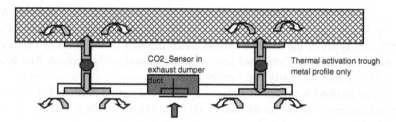

Fig. 20 Ceiling panel for heating and cooling, with integrated extract air terminal and lighting

In this solution, the activation of the thermal storage is through profiles/pipes fitted with hydraulic circuits spreading heat in both directions: on the surface of the panel for a direct exchange and on the concrete surface to activate the inertia of the concrete flagstone, Fig. 20.

For efficient operation, 50 % of the flagstone had to remain as raw concrete. Through the activation of the flagstone, the peak power was reduced by about 35 %. This had a positive impact on the design of plants and helped to control investment costs.

3.4 Automation of the Next Generation

At the level of the building's automation, a new technology has found its application. The management of decentralized units and the recovered air dampers and CO_2 sensors are driven by Digitalstrom. This technique uses the electric power for the transmission of information and makes the installation of a conventional bus obsolete. Given that the implementation of this system was a world first and it was necessary to consider some "teething troubles", the system was limited to the installation of ventilation of the offices. For all other HVAC systems, lighting and blinds, a traditional LON system has been implemented.

3.5 Maximized Daylighting

The lighting concept supports the use of daylight. Each workplace is located in front of a large bay window. The windows are generously sized, with 3.2 m^2 per workplace. The depth of the premises is 5 m only. Most of the time, work in daylight is thus possible. Thanks to architectural measures, workplaces are protected from the solar glare. As light source, fluorescent tubes of TL5 type, with reduced mercury levels, were installed. The concept has been supplemented by LEDs. In offices, fluorescent tubes are used as basic lighting and LED table lamps serve as support lamp at the level of workplaces. In the corridors, the LEDs are used as decorative lighting and create a pleasant atmosphere. The meeting rooms are equipped with HIT lamps complemented by LED lights for atmosphere. With this combination of lighting, the specific power amounts to only 6.6 W/m². Only six different types of light fixtures are installed throughout the building enclosure, and thus, maintenance costs and servicing are considerably limited. Movement detectors and light intensity can further reduce consumption to a minimum.

Exterior lighting was kept to a minimum to avoid light pollution. The idea of a night enhancement of the façade and of the illumination of the natural garden has been abandoned.

3.6 Efficient Water Management

A system of efficient water management is based on three axes: the first is the reduced consumption of drinking water because it contains a significant proportion of energy for transportation and treatment. Drinking water is distributed in the kitchen for water fountains located in the building as well as in the washbasins of the lavatory. The second axis is to use grey water for toilet flushing and garden irrigation. This water, collected on the roof, is carried in a tank of 70 m³. The overflow is led directly to the natural garden and, in fact, to the groundwater. The last axis is the optimization of drinking water. Water flows in the taps were limited, and the taps in the toilets were equipped with infrared detectors. Result of this concept: a saving of water of 4,000 l/day (Fig. 21).

3.7 First Performance Review After 8 Months

As the real optimization phase has not begun jet, the only figures available today are the total electricity consumption for the new building. Compared to the dynamic building simulation including all electrical energy consuming facilities,

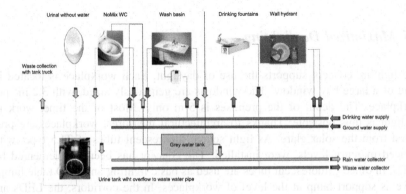

Fig. 21 Principle of water management

Table 7 Simulated energy performance of the building

	Delivered and exported energy, kWh/(m² a)	Primary energy factor	Primary energy, kWh/(m² a)
Space, water and supply air heating, electricity to heat pumps	6.0	2	12.0
Cooling, electricity to heat pumps	6.7	2	13.4
Fans (HVAC)	5.3	2	10.5
Pumps (HVAC)	2.8	2	5.6
Lighting	16.3	2	32.6
Appliances (plug loads)	26.8	2	53.6
PV power generation	−30.9	2	−61.8
Total	**33**		**66**

All specific values are per net floor area

the one-year consumption hit the simulation target quiet well. Breakdown of the simulated energy performance is shown in Table 7.

For the first year, the calculated values will overrun the calculated electrical consumption by 10 %, Fig. 22. This might be optimistic because the building is today occupied by 90 %. In addition to that, some troubles with the ventilation control system have been fixed during the last month. Focused on this early result, the analysis shows big discrepancies during October and November. Further investigations are necessary to improve the whole system. The goals for the next step is to break down the comparative results and analyse consumer one by one to check whether set points, operation schedules and the sensor technique are running correctly. Also user behaviour need to be analysed.

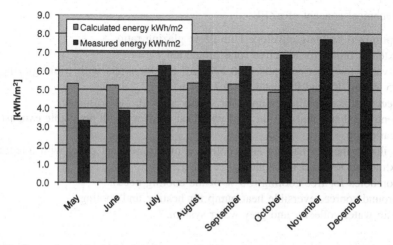

Fig. 22 Comparison of calculated and measured delivered energy use

Even with the analysis not finished jet, the result of energy performance for the IUCN extensions building proves that the annual energy use is able to hit the MinergieP target.

In a general way, the building designed according to Minergie Standard shows a coherent behaviour between design parameter and real measurements. It is valid for peak power demand for heating and cooling and lighting power. On the other hand, the calculated energy use does not fit exactly to the real building consumption. The main reason for that is that standard calculation does not correspond to real behaviour and occupation scheduled. Variation from $\pm 20\%$ can be expected. Important for high-performance buildings is that the variation between calculated and real measures varies in the same percentage range as in normal buildings. Based on a low net energy need for heating of 22 kWh/m²/year, the result can vary of 4.4 kWh/m². On this low level of energy consumption, it is more than comprehensive that the user behaviour has a higher impact than on normal buildings. In general, MinergieP buildings have kept their premises in terms of energy savings if used as designed.

3.8 Key Facts of the IUCN Headquarter Building

Floor area and volume:

Net floor area	4,530 m²
Volume of building according to SIA 116	31,700 m³
Volume of building according to SIA 416	26,115 m³

Technical concept of nZEB:

- Optimized building envelope with 25 % glazing ratio.
- External blinds for effective solar protection.
- U-value for exterior walls of 0.1 W/(m² K), for walls with triple glazing of 0.5 W/(m² K) and for windows 0.7 W/(m² K).
- Decentralized ventilation units for supply air with facade intakes.
- Central exhaust units on the roof with heat recovery with reversible exhaust air heat pump.
- Ceiling panel for heating and cooling, a multifunctional panel with integrated extract air terminal and lighting.
- Boreholes for free cooling (30 % of the cooling need).
- Ground-source reversible heat pump for heating and cooling.
- Rain water collector and grey water system.

Energy data:

Annual total electricity use (including user appliances)	289 MWh
PV installed power	150 KWp
Electricity generated with PV system (calculated)	140 MWh/a
Delivered en. (all electricity, including user appliances)	64 kWh/(m^2 a)
On-site electrical energy production with PV	31 kWh/(m^2 a)
Delivered minus exported energy	**33 kWh/(m^2 a)**
Primary energy	**66 kWh/(m^2 a)**
Saving of drinking water about	4,000 l/day

4 TNT Green Office in Hoofddorp, Holland

The design of the TNT Green Office is characterized by sustainability, transparency and connectivity. First, a volume study was done to test different volumes regarding criteria such as compactness, flexibility, daylight factor, view, building costs and the highest LEED results. The design consists of two rectangular parallel volumes, each six stories high. On the Westside (the Geniedijk), the lower three stories of these two volumes are connected by terrace-like volumes and the upper three stories by connective bridges. These connections offer great meeting places for the employees as well. On the Eastside, both volumes are connected by a third "floating" volume. For the TNT headquarters, 16-m-long concrete floor slabs were used that were made out of recycled rubble and granulate. Due to the long span, fewer supports were needed, which saves material and generates spaces that can be divided up freely (Fig. 23).

Fig. 23 Facade of the TNT Green office building Hoofddorp

4.1 General Building Information

Building type: office building	
Owner: TNT group	
Location: Hoofddorp, Holland	
Key building parameters:	
Number of storeys	6 levels
Net floor area	16,136 m^2
Gross floor area	17,956 m^2
Floor area parking garage	7,207 m^2
Number of occupants	873
Construction year: 2011	
LEED	Platinum certificate
Design and construction:	
Design	Architectenbureau Paul de Ruiter b.v.
Building physics	DGMR
Structural design	Van Rossum Raadgevende Adviseurs
Building Services	Deerns, Rijswijk
Consultant LEED	B en R Adviseurs voor duurzaamheid
Main contractor	Boele & van Eesteren
Facade	De Groot & Visser
Building services contractor	Kropman
Process control	Schneider Electric

4.2 Massing and Daylight

The atrium has been designed in such a way that as many daylight as possible can enter and it offers the employees a beautiful view. The atrium and the entrance are clearly connected, and the terrace-like volumes encourage employees to take the stairs instead of the elevator, thereby serving both a health and social purpose (Fig. 24).

The presence of daylight in living and working areas is of crucial importance to the well-being of the working and living environment and also for the health of the user. Daylight was the leitmotif in the design of the building, which has a completely glazed north façade. The Schüco mullion-transom FW 50+ was used as a façade system, offering a very narrow profile face widths and large module widths, as well as excellent sound and thermal insulation. Due to their narrow face widths, the window systems aluminium window system (AWS) 102 and AWS 65 allow for a large amount of glazing. U-value of the glazing units is 1.4 W/(m^2 K), and the solar heat gain coefficient g is 0.27 or 0.33, depending on facade, corresponding to visible transmittance of 0.5 and 0.6, respectively. Intelligent solar shading louvers were installed on the façade.

Fig. 24 Roof and area of the atrium

4.3 Heat Load and Energy

Goal was to minimize the heating load as much as economical possible made sense, high levels of insulation were applied:

Window U-value	1.4 W/(m^2 K)
Interior wall U-value	3.3 W/(m^2 K)
Floor on ground U-value	0.24 W/(m^2 K)
Floor above garage	0.32 W/(m^2 K)
Internal ceiling	4.2 W/(m^2 K)
Roof U-value	0.24 W/(m^2 K)

Also, the internal heat gains were reduced as much as possible to reduce the cooling load in summer. The different equipment types included in the calculation of miscellaneous equipment are as follows: PC's, mLCD monitors, printers/copiers/scanners, communication and A/V, and kitchen and restaurant equipment. The data centre equipment comprises the computer server equipment located in the main equipment room and satellite equipment rooms and spread throughout the building. This led to the following results: office rooms 28 W/m^2, from which persons and equipment 20 W/m^2 and lighting 8 W/m^2, total energy use office appliances 19.2 kWh/(m^2 a) and total energy use of data centre equipment 24.0 kWh/(m^2 a). The breakdown of energy use is shown in Table 8.

Maximum heating need	
Air handling	28.6 W/m^2
Transmission	14.2 W/m^2
Infiltration	8.5 W/m^2
Total	51 W/m^2
Maximum cooling need	
Air handling	13.1 W/m^2
Cooling load in rooms	41.8 W/m^2
Total	55 W/m^2

Table 8 Simulated energy performance of the building

	Delivered and exported energy, kWh/(m^2 a)	Primary energy factor	Primary energy, kWh/(m^2 a)
Heating, electricity to heat pumps	9.8	2	19.6
Hot water, electric boiler	3.5	2	7.1
Cooling, electricity to heat pumps	3.3	2	6.6
Fans	16.8	2	33.7
Pumps	0.7	2	1.5
Elevators	0.8	2	1.5
Lighting (interior)	21.1	2	42.2
Lighting (exterior)	0.8	2	1.6
Appliances (plug loads)	19.2	2	38.3
BioCHP electricity generation	−73.8	2	−148
BioCHP fuel consumption	184	0.5	92.2
Heating energy exported to other buildings (estimated value)	−50	0.5	−25
Total	**137**		**72**
Not included in the building energy balance			
Data centre electricity	24.0	2	48.0
Heat rejection of BioCHP coolers (electricity)	10.0	2	20.0

All specific values are per net floor area. The data centre electricity use and heat rejection of bioCHP coolers are not included in the building energy balance

4.4 Ventilation

The overall ventilation for the building is done by four central air handling units, all equipped with heat recovery systems. HVAC total supply air volume is 111.091 m^3/h, static fan pressure 944 Pa for supply air fan and 688 Pa for extract fan, with total fan energy use of 16.8 kWh/(m^2 a) (Fig. 25).

4.5 Divers Sustainable Measures

Several sustainable techniques are applied: intelligent awning, hybrid ventilation (natural if possible, mechanical if necessary), heat recovery from the extract air, energy-saving equipment and lighting, long-term cold/heat storage in the aquifer, on-site generation of electricity through the use of bioCHP (Green Machine) and an advanced building management system.

Fig. 25 Ceiling panels for heating and cooling with integrated air diffusers

4.6 Long-Term Energy Storage

The surplus heat of heat pumps in the summer and the surplus of cold in the winter are stored below ground level in the aquifer. The stored heat is used through heat pumps to warm the building in the winter, and the stored cold to cool it down in the summer. The electricity for the two heat pumps of 332 kW is delivered by the bioCHP.

Key design parameters of long-term energy storage system:

Power LTEO	715 kW
Temperature difference	8 K
Groundwater flow	77 m³/h
Reserve	20 %
Groundwater flow including reserve	92 m³/h

4.7 BioCHP

All electricity for the Green Office (on yearly basis) is generated on-site in a sustainable way using a bioCHP. For the remainder of the peak demand, green electricity is purchased. This way, the TNT Green Office operates completely CO_2 emission free. The produced heat of bioCHP is supplied to nearby (yet to be realized) office buildings. During the periods when bioCHP heat production cannot be fully utilized, the excess heat is rejected with roof placed coolers.

The bioCHP installed is a Cummins KTA 19 bioCHP unit of electric power 300 kW with a Stamford HCI 534C generator, which should generate 1.200 MWh/a [73.8 kWh/(m^2 a)]. The bioCHP unit is running on purchased biodiesel (palm oil) and generates electricity for the building. The electrical efficiency of bioCHP plant is 40 % and the total fuel efficiency with heat production 86 %.

4.8 Solar Hot Water Heater

A solar hot water heating system has been added to the building. The system includes two solar collectors with size of 2.4 m^2 each. The system contributes heat towards the DHW system. A total of 0.25 kWh/(m^2 a) of heat is collected on average each year. Since the DHW system comprises electric boilers, the energy contribution of the solar collectors displaces electric energy.

4.9 Results GreenCalc+ and LEED

Based on the design and the actual realization, the Milieu Index Gebouw has been calculated for the TNT Green Office, the building will have an Environmental Index of at least 1,000 points in accordance with the GreenCalc methodology. This is more than 1.5 times better than the current building with the highest Building Environmental Index. This index is determined on the basis of the materials and the quantities used. During the design and preparation for the construction of the building, continuous attention is devoted to assessing whether the choice of materials is the most environment-friendly and whether the quantities remain within estimates so as to guarantee that the building achieves an ultimate index of at least 1,000. Not only materials count towards this goal, also how water and energy are treated contribute to a higher index. The TNT Green Office has a bioCHP (combined heat and power production from biofuel) for the purpose of generating power. The high score for energy is due to the compensation effect of applying the bioCHP, without that the score for energy would have been around 220. This would lead to a GreenCalc+ score of 481, still among the best, for example one of the most environment-friendly office buildings in the Netherlands, the 2004 Rijkswaterstaat building in Terneuzen, has a score of 323 (Table 9).

Table 9 Environmental index calculation with the GreenCalc methodology

	Material cost a year		Milieu Index
	Design	Reference	MIG
Materials	€28.473	€38.352	135
Energy	€0	€264.958	∞/200
Water	€1.138	€1.663	146
Total	€29.610	€304.973	1,030/481

The LEED assessment takes place in the area of design, implementation, the ultimate use and management. This is tracked in five categories: materials, energy consumption, efficient water use, interior environment and the environment. By way of example: the location is important, therefore including the proximity of the building to the public transportation network. The methodology even extends to the need for documenting the specific properties of the paints used. These may not contain more harmful substances than prescribed by LEED. In addition, a pre-scribed minimum quantity of recycled materials must be used and a large per-centage of the materials used must be "regional". The highest certificate that can be issued under this methodology is the LEED Platinum Certificate and this will indeed be the certificate awarded to the building.

5 Hagaporten III in Solna, Sweden (Fig. 26)

General building information:

- Building type: office building
- Owner: Norrporten, developed by Skanska 2008
- Key building parameters:
  - Number of storeys   7 levels above ground + 3 levels below ground
  - Height   30 m
  - Net floor area   30,000 m² office area
  - Gross floor area   20,000 m² parking area
  - Occupant density   15 m²/person (overall average)
- Construction year   2008
- Construction costs   61 million €

Fig. 26 Hagaporten III office building

Outdoor climate:

• Design outdoor temperature for heating	−20 °C
• Design outdoor temperature and RH for cooling	27 °C/50 %

Indoor environmental quality targets:

• Indoor air quality	
• Airflow rate	45 m³/h per person
• Indoor temperature	22–23 °C (20–25 °C)
• Illuminance level	300 lx

Heating and cooling load, and building envelope:

• Cooling load	40 W/m²
• Heating load	40 W/m²
• Building envelope:	
– Window *U*-value	1.4 W/(m² K)
– Window *g*-value	0.4
– Exterior wall *U*-value	0.32 W/(m² K)
– Roof *U*-value	0.13–0.15 W/(m² K)
– Building leakage rate at 50 Pa	0.5 l/(s m²)

Environmental targets:

- EU Green Building Label;
- Miljöbyggnad Label (Swedish)
- Carbon footprint 4.5 kg CO_2/(m^2 a) (half the amount of average office building in Sweden)
- Environmental Management Standard ISO 14001.

Energy performance targets:

• Delivered energy target kWh/(m^2 a):	85
– Heating (district heating)	39
– Cooling (district cooling)	26
– Electricity (excluding lighting and tenant power)	20

5.1 Energy Sources

District heating is generated from treated wastewater and hot water boilers, which are primarily fuelled by wood briquettes and pine oil. District cooling is generated from chillers and cool water from the Baltic Sea.

5.2 HVAC Technologies

The ventilation system far exceeds the air exchange requirements of the Swedish building code and ensures good indoor air quality. The ventilation system is equipped with a low-speed high-efficiency air handling unit, free cooling that returns heat from office back to office and an efficient heat recovery system that recycles approximately 70 % of the heat energy from the exhaust air. Additionally, the building is heated and cooled by district heating and cooling systems, which are more efficient than localized systems.

A CAV system comprising of three AHUs supplies the offices. Each AHU provides 13 m^3/s supply air and is fitted with a "free cooling coil" connected to the chilled beam circuits. The heat recovery temperature efficiency of this coil is 69 %. Each AHU has a specific fan performance (SFP) of 1.4 kW/m^3. The low air speed of 1.6 m/s through the AHU means that no sound attenuators are needed in either the AHU or the ducting system (Fig. 27).

One AHU for restaurant/kitchen area of 8 m^3/s is equipped with heat recovery. The kitchen hood is equipped with ozone lamps for deodorizing the exhaust airflow. Exhaust air from the offices is transferred to ventilate and heat the garage before leaving the building through the garage exhaust AHU fitted with a heat

Fig. 27 Free cooling coil in AHU cools the chilled water system and preheats the supply air

recovery coil. The garage extract AHU uses run around coils to reclaim the heat from the exhaust air. Both the supply and exhaust systems are operated at a constant pressure 100 Pa by controlling the fan speed. There are no balancing dampers in the ducting which is maintained at a constant static pressure relative to the fan speed; only the final pressure drops in the supply air diffusers, and active chilled beam ensures the correct balanced air volumes at each supply air outlet.

During the evening and night-time, the ventilation can be started manually on reduced speed for overtime ventilation, by a switch placed at each staircase on each office floor.

The chilled beam chilled water system is also operated at a constant pressure of 30 kPa by speed control of the pumps. Summertime district cooling provides cooling. In winter time, all chilled water to the active chilled beams is provided by the coil in the AHU air intake, which both cools the chilled water and preheats the fresh air. The internal heat load in the chilled beam chilled water system does not leave the building but reused by the ventilation system by preheating the supply air.

The following considerations were used in the design of BMS:

- Web-based control system in technical rooms;
- No bus system on office floors. Local control units for forced ventilation in meeting rooms;
- BMS equipped with remote energy monitoring and multiple data points.

5.3 Daylight and Electric Lighting

Glass facades and an inner daylight well at the centre of the building allow natural light into the building and to all workstations.

The windows are highly insulated to minimize heat loss during cold weather and are tinted to reflect approximately 85 % of the radiation from the sun and reduce solar heat gain.

Energy-efficient lifts and low-energy lighting have been installed and occupancy sensors control lighting in spaces not regularly occupied.

5.4 Functional and Flexible Building

Hagaporten 3 is a functional building capable of fulfilling all modern office building requirements and has been designed to be flexible to accommodate the future needs of tenants and to promote a long useful lifespan. Each floor has the potential to be occupied by one single company, or many smaller companies with individual tenant entrances, and the office spaces are designed to allow tenants to customize and create their own unique space identity. The interior design also enables easy rebuilding to meet the future needs of tenants. The ventilation system, which circulates large volumes of air throughout the building, enables the future expansion of the building without significant upgrade to the existing system.

5.5 Other Sustainability Considerations

Low-emission building materials. Low-toxicity materials were used within the building to ensure the total volatile organic compound (VOC) levels that were less than 200 $\mu g/m^3$, compared to the Swedish standard of 300 $\mu g/m^3$. The flooring and ceiling were made with natural materials and wood oils and paints approved by the asthma and allergy association have been used to minimize indoor pollution and avoid adverse impacts on human health.

Reduced external noise. The walls, windows and glass facades are designed to reduce external noise disturbance from the nearby E4 motorway.

Electromagnetic radiation protection. Electrical junction boxes have been distanced from regularly occupied areas to ensure that electromagnetic radiation will not exceed 10 V/m.

Recycling facilities. Hagaporten 3 is equipped with recycling facilities for office waste, which can be customized to meet specific tenant requirements.

Green roofing. The building has a sedum vegetation roof, which is drought resistant and reduces runoff during wet weather by absorbing water. The roof also provides additional insulation, creates wildlife habitats, absorbs atmospheric pollution and extends the lifespan of the roof by protecting the roof surface from UV light.

5.6 Measured Performance

Measured energy performance compared to simulated energy performance is shown in Table 10. Lighting and tenant's small power loads have not been included in the design target in this building. Primary energy has not been possible to calculate, because primary energy factors are not in use in Sweden. However, because of dominating renewable sources, non-renewable primary energy factors at least for district heating and district cooling will be reasonably low.

The envelope performance in terms of building leakage rate per building envelope square metre at 50 Pa pressure difference (between outdoor and indoor) was tested with the result of 0.5 l/(s m^2) corresponding exactly with the target.

5.7 Compliance with Environmental Labels

Healthy indoor office environment. The use of natural light, high-quality ventilation, high use of non-toxic construction materials, sound insulation and minimized electromagnetic radiation levels have contributed towards a healthy working environment for the tenants of Hagaporten 3.

Urban redevelopment and planning. The site was previously used as a car park, car wash and vehicle test-drive area. Hagaporten 3 is thought to have improved the perception of the area from the E4 motorway, which is a major gateway to central Stockholm. The building also has reduced levels of noise, dust and pollution from the E4 motorway in adjacent residential areas by acting as a barrier.

Hagaporten 3 is within walking distance of amenities, shops and restaurants in the Solna area. The building has good access to public transport, including several city bus routes, the Stockholm Arlanda airport bus route and the commuter train from Solna station, which is 10 min from Stockholm's Central Station. Central

Table 10 Measured and simulated energy performance. The results are reported as delivered energy, because primary energy factors are currently not available in Sweden

	Design target	Measured	
	Delivered energy, kWh/ (m^2 a)	Delivered energy, kWh/ (m^2 a)	Energy carrier
Space, hot water and supply air heating	39	43	District heating
Cooling	26	18	District cooling
Fans and pumps	20	17	Electricity
Total	**85**	**79**	
Not included in the design target			
Tenant's electricity (appliances) and lighting		53	Electricity
Process cooling (refrigerated cabinets, etc.)		12	District cooling

Stockholm is approximately 15 min by bicycle and the building is equipped with showers, and indoor and outdoor cycle storage areas for cyclists.

Raising awareness of more sustainable buildings and construction. Skanska are raising awareness of more sustainable construction through their partnership with the EU Green Building programme. It is hoped that the programme will influence the construction industry to invest in more energy-efficient buildings. During the project, communication with Solna Council is thought to have led to a greater awareness and understanding of energy-efficient buildings and environmental construction methods within the council.

Construction employment. Between 160 and 180 workers, mostly subcontractors worked on the site throughout the construction phase. Approximately 90 % of the workers lived within commuting distance of the site.

Vocational training. Training in manoeuvring scaffolding and platforms was provided during the project, according to new Swedish regulations issued in 2007. Two workers also received training and obtained their forklift driving licences.

Minimizing environmental impacts during construction. Noise disturbance, dust and air pollution were minimized with consideration to the surrounding residential and commercial areas. Noise was monitored prior to and during heavy construction activities, such as excavating, rock blasting and piling, to ensure that national and municipal noise regulations were not exceeded. Construction roads were treated with dust-binder during dry periods and industrial vacuum dust collectors were used during drilling to reduce airborne dust pollution. Environmental class one fuels were used for site machinery, petrol motors smaller than 20 kW, used cleaner fuels and catalytic converters were fitted, where possible, to minimize air pollution.

Environmental programme. The environmental programme ensured that the project was compliant with the International Environmental Management Standard ISO 14001 and that all local and national regulations were adhered to. Construction waste was minimized, site drainage issues addressed, trees and plants were preserved where possible and surrounding land was not damaged during construction.

Less than 15 % of construction waste went to landfill, which reduced project costs as well as the overall environmental impact. Waste was sorted on-site and recycled externally, and combustible non-recyclable waste was used at a district heating plant.

Environment-friendly construction materials. The use of potentially hazardous construction materials was restricted, and all materials were checked against Skanska Sweden's chemical rules and the BASTA system, which is an agreement within the Swedish construction industry to phase out substances hazardous to the environment and human health. As a consequence of BASTA compliance, non-halogen lighting containing minimal mercury and no bromide flame retardants were used within Hagaporten 3. Environmentally certified materials included the flooring and sections of the roof.

5.8 Experience Gained from the Operation

When the building was constructed and operational, a separate "energy related" commissioning of the AHU was made. In consequence, adjusting the brine water flows in the run around coil increased efficiency by approximately 5 %, and re-balancing of the brine circuit by reducing the pump speed and opening up the balancing valves reduced the power consumption of the brine pumps by more than half. During the first months after completion, energy consumption was still too high which was identified by the energy follow-up procedure making an energy signature for the buildings different operating modes. The energy signature was then combined with a climate file consisting of hourly recording of the external temperature, sun radiation, etc. The resulting calculations showed that the measured annual energy consumption was too high. Subsequently numerous different actions were undertaken, for example reprogramming the control of the chilled water system pumps and valves not to allow the chilled beam system to operate at night and weekends without the AHU saved approximately 5 kWh/m^2 a. Reducing the radiator system temperature during the night saved a further 8 kWh/m^2 a. After one-year operation, there are still some fine-tuning improvements to be made in the annual energy use, but the priority is now to ensure a continuously good energy performance of the building (Fig. 28).

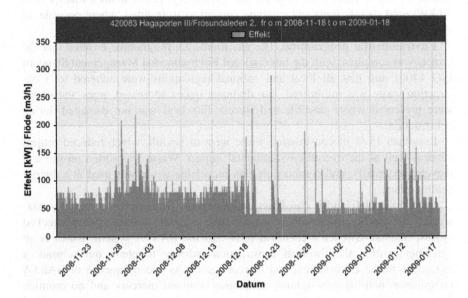

Fig. 28 The reduced cooling demand is due to no pump or valve control operation in night-time for comfort, reprogramming made 2008-12-18. The spikes above were removed later by preventing the control valve from opening before 1 h of the AHU's operation each morning. Instead, free cooling cooled down the chilled beam circuit in less than 1 h

6 Summary of nZEB Case Studies

Reported nZEB case studies show that some solutions and principles have been common in all buildings. Dependency on the climate and the possibility to use many alternative solutions can be also seen. A common design principle, to start with demand reduction measures that are then supported with effective HVAC systems and on-site renewable energy production, was used in all buildings:

$$nZEB = \text{demand reduction} + \text{effective HVAC} - \text{systems} + \text{on-site renewables}.$$

Energy sources used depended very much on the building site. If district heating and cooling were available, they were also used. Different types of heat pumps were popular, and in one case, bioCHP was used. The following technical solutions, commonly used in all case study nZEB buildings, can be listed:

- Balanced heat recovery ventilation, i.e. mechanical supply and exhaust ventilation with heat recovery, called also dedicated outdoor air system (DOAS), was used in all buildings. Ventilation system was centralized in four buildings and decentralized without supply air ductwork in one building in Gland. Demand-controlled ventilation (at least in some rooms) or very low pressure systems were used to reduce the fan energy. Mechanical ventilation was combined with natural stack effect ventilation for ventilative cooling purposes in Dion and Helsinki. In Dion, the system served all office areas and was boosted with central exhaust fan; in Helsinki, the system was limited to atrium spaces.
- Free cooling solutions were typically combined with mechanical cooling. For free cooling, boreholes were used directly or with water to water heat pump, evaporative and ventilative cooling was also used. In Helsinki, free cooling directly from boreholes provided 100 % of cooling (i.e. in this building, there was no mechanical cooling installed), and in Solna, district cooling was used.
- Optimized facades and effective external solar shading were used in most of case studies as well as the utilization of thermal mass and other passive measures were typical.
- Utilization of daylight with occupancy sensors and photocell-controlled dimming was a common solution for electric lighting in these buildings, where both fluorescent and LED lamps were used.
- High-efficiency components and distribution systems were commonly used. This applied for high heat recovery efficiency, low-specific fan power, and water-based distribution systems for heating and cooling.
- Solar PV was the most common on-site renewable energy production solution.

With all these common measures applied in nZEB case study buildings, energy performance was improved remarkable compared to conventional modern office buildings. Energy uses of heating, ventilation, cooling and lighting were so well controlled that office appliances became the major component in the energy balance of these nZEB buildings. Overall primary energy (including office appliances) was

Table 11 Comparison of climate depending solutions based on case study buildings

Central Europe	North Europe
Larger windows for max daylight to save lighting electricity	*Small windows* for lowest acceptable average daylight factor
Moderate insulation ($U_{window} = 1.1$, $U_{wall} = 0.30$)	*Highly insulated envelope* ($U_{window} = 0.6...0.8$, $U_{wall} = 0.15$)
More cooling need than heating need	Slightly less cooling but much more heating
External solar shading	External shading for large windows
"Glass" buildings with external shading possible	*Double skin façade* to be used for "glass" buildings
Free cooling combined with compressor cooling	100 % free cooling possible with borehole water
Water-based distribution system for cooling	Water-based distribution systems for *heating and cooling*

between 57 and 85 kWh/(m² a). The highest value of 85 kWh/(m² a) was from Helsinki, where heating clearly dominated despite a lot of demand reduction effort, showing that a cold climate increases energy use.

As two of these buildings were from North Europe and three from Central Europe, climate depending solutions were possible to see, Table 11.

Windows were larger in warmer climate because of better daylight utilization and less critical heat losses. Large windows also need external solar shading, which can be avoided in a cold climate, if minimum size windows just providing required daylight are used to control heat losses. The latter is important, because in a cold climate heating dominated, by a factor of almost 10 compared to Central Europe. In Central European buildings, quite different insulation level was called as "highly insulated". The best insulation and triple glazed windows were used in Gland (other two Central European buildings were with double glazing). Therefore, typical values shown in Table 11 might need some improvement at colder parts of Central Europe.

Design process description of these buildings showed that integrated design (similar to the approach described in Chap. 7) was used in all cases to achieve challenging targets that has been important in both climates.

Printed in the United States
By Bookmasters